Springer Theses

Recognizing Outstanding Ph.D. Research

For further volumes:
http://www.springer.com/series/8790

Aims and Scope

The series "Springer Theses" brings together a selection of the very best Ph.D. theses from around the world and across the physical sciences. Nominated and endorsed by two recognized specialists, each published volume has been selected for its scientific excellence and the high impact of its contents for the pertinent field of research. For greater accessibility to non-specialists, the published versions include an extended introduction, as well as a foreword by the student's supervisor explaining the special relevance of the work for the field. As a whole, the series will provide a valuable resource both for newcomers to the research fields described, and for other scientists seeking detailed background information on special questions. Finally, it provides an accredited documentation of the valuable contributions made by today's younger generation of scientists.

Theses are accepted into the series by invited nomination only and must fulfill all of the following criteria

- They must be written in good English.
- The topic of should fall within the confines of Chemistry, Physics and related interdisciplinary fields such as Materials, Nanoscience, Chemical Engineering, Complex Systems and Biophysics.
- The work reported in the thesis must represent a significant scientific advance.
- If the thesis includes previously published material, permission to reproduce this must be gained from the respective copyright holder.
- They must have been examined and passed during the 12 months prior to nomination.
- Each thesis should include a foreword by the supervisor outlining the significance of its content.
- The theses should have a clearly defined structure including an introduction accessible to scientists not expert in that particular field.

Hannu Christian Wichterich

Entanglement Between Noncomplementary Parts of Many-Body Systems

Doctoral Thesis accepted by
University College London, United Kingdom

 Springer

Author
Dr. Hannu Christian Wichterich
Department of Physics and Astronomy
University College London
Gower Street
WC1E 6BT London
United Kingdom
e-mail: hannu@theory.phys.ucl.ac.uk

Supervisor
Prof. Sougato Bose
Department of Physics and Astronomy
University College London
Gower Street
WC1E 6BT London
United Kingdom
e-mail: s.bose@ucl.ac.uk

ISSN 2190-5053   e-ISSN 2190-5061

ISBN 978-3-642-19341-5   e-ISBN 978-3-642-19342-2

DOI 10.1007/978-3-642-19342-2

Springer Heidelberg Dordrecht London New York

Cover design: eStudio Calamar, Berlin/Figueres

Printed on acid-free paper

Springer is part of Springer Science+Business Media (www.springer.com)

Supervisor's Foreword

Entanglement is the name for the most peculiar correlations that can exist solely between quantum mechanical objects. These correlations become inexplicable when one wants to ascribe precise states to individual objects and were originally thought to signal an incompleteness of quantum mechanics. Yet, the correlations are so strong and weird that they have become a prime resource for practical applications, for example, in the new field of quantum information processing. Not only can they be used for teleporting an arbitrary quantum state and establishing a secret random key for cryptography, whose security is guaranteed by the laws of nature, they can also enable full scale quantum computation aided solely by measurements. Moreover, entanglement is a quantifiable resource, which makes it sensible to ask the question of as to "how much" entanglement exists in the systems that surround us.

A fruitful arena to look for entanglement is inside strongly correlated many-body systems. It has long been appreciated that the lowest energy configurations of such systems can only be reached if one allows for entanglement between their constituents. However, investigations into the amount and type of entanglement in many-body systems are comparatively a new line of research. It is in this area that the thesis of HannuWichterich breaks new ground. Previously entanglement had been studied between individual components of a strongly correlated system or, at the other extreme, between its complementary parts. The latter was found to be so large as to diverge with the size of the system at the so called quantum critical points. An open question remained as to what was the entanglement between separated blocks of such a system. Apart from the fundamental curiosity, this question is motivated by the fact that it is the entanglement between well separated systems which is a resource for quantum information applications. Not only was this an uncharted territory so far, its quantification necessitated the use of a difficult to compute measure of entanglement. It is here that HannuWichterich used numerical techniques in a clever way to compute, for the first time, the entanglement of separated (i.e., non-complementary) blocks of one dimensional strongly correlated spin systems. This computation and its findings, namely a "scale invariance" of entanglement at quantum critical points, and a unification of a few disparate types of entanglement under one formula, form an interesting part of the thesis. The work attracted wide attention of quantum field theorists investigating other information measures between non-complementary parts of a many-body system and received several citations from them. Wichterich's methods and this form of entanglement are both likely to stimulate much further research and indeed the former have already been successfully used by other members of our group to gain insight into many-body systems, such as Kondo systems. In his thesis, Wichterich also describes an analytic investigation of the same type of entanglement using infinite range many-body models, which complements the above work to show that some analytic investigations may indeed be possible for appropriate models.

Another interesting topic discussed in the thesis is the ability to extract entanglement in its purest, most useful, form by means of certain coarse grained measurements on distant parts of a many–body system. This work introduces a completely new notion of entanglement which may be computed in various instances. However, from a practical application point of view, the most important part of the thesis is the study of non-equilibrium dynamics to yield long distance entanglement. Non-equilibrium dynamics of many–body systems is currently an issue of high topical interest, and usually studied to investigate the equilibration of such systems and the production of defects through rapid quenches. Wichterich reports a completely new application of dynamics following quenches, by showing that the farthest spins of an open ended spin chain may be entangled by this method. The amount of the entanglement can be so high even for significantly long chains that after some post processing, it can be useful for quantum teleportation between distant registers of quantum computers. This work has also had a signicant following with many other authors investigating quenches with a similar motivation.

The thesis will be most helpful for researchers aiming to start working at the interface of quantum information and many–body theory. While the introduction will bring them up to date on what has been done in the area, the original parts of the thesis will provide them with fresh angles to launch studies at the above interface. The contents of the thesis are well supplemented with pertinent appendices which will very fruitfully add to the quantum information knowledge of many–body theory theorists and provide some necessary many–body techniques for the quantum information specialist.

London, January 2011   Sougato Bose

Acknowledgments

I am indebted to a number of people who in the course of my PhD studies were of invaluable support and assistance.

First and foremost, I would like to dearly thank my supervisor, Prof. Sougato Bose, who offered me a PhD position in his QUINFO group at UCL and who has since been a valued source of inspiration and advice. I owe him my deepest gratitude for his devoted supervision, and for sharing his broad expertise and physical intuition which has been pivotal for the efficient progress of my work. I thank my examiners, Prof. Tania Monteiro and Dr. Terence Rudolph, for an enjoyable PhD viva and for a number of valuable suggestions which helped me improve this thesis. I would also like to show my gratitude to Prof. Jonathan Tennyson, who nominated my thesis for publication in the series Springer Theses.

For being a good friend and an ambitious and persistent collaborator, and for the numerous contributions to this work let me dearly thank Javier Molina, who despite my suspicious nature always kept the faith with our endeavours.

I am also indepted to my collaborators Julien Vidal, Pasquale Sodano, and Vladimir Korepin for their contributions to different projects of my PhD studies and for sharing their wealth of expertise.

Moreover, let me thank Marcus Cramer, Abolfazl Bayat, Janet Anders, Alessio Serani, Jochen Gemmer, Ingo Peschel, Viktor Eisler, and Pasquale Calabrese for vastly helpful discussions.

Let me extend my thanks to the whole QUINFO group who made my time at UCL so very enjoyable.

For being a good friend, for an excellent proof-reading of this thesis, and for helpful advice on countless occasions I would like to thank Tom Boness. I will always remember the enjoyable time we shared as PhD students at UCL.

Without the love, trust, and support of my parents, Rolf and Dörte Wichterich, this work would not have been possible. For being great parents and for being supportive in every way imaginable: Thank you!

My dearest, loveliest Eva. For enduring my occasional mood swings into grumpiness, for being there for me and sharing the best time of my life with me I thank you from the $\heartsuit$.

Contents

Chapter 1
Introduction

Ground states and out-of-equilibrium states of many-body systems (MBS) usually carry a large amount of correlations among the degrees of freedom of their constituents (Fig 1.1). Without correlations such a collection of particles would simply behave as the sum of its parts, but emergent phenomena such as the different phases of matter could not be explained under this premise. While classical correlations have been a long standing topic in the characterisation of phases of matter, the role of entanglement—the purely quantum part of correlations—in these phenomena has attracted the attention of theorists and experimentalists during the past decade [1].

On the one hand, entanglement in quantum states renders simulation of MBS a hard task [2]. On the other hand, its study reveals new aspects of the theory of strongly correlated systems, for instance it may serve as a diagnostic of quantum phase transitions [3]. The study of entanglement in many-body states is also fuelled by recent experiments on cold atoms trapped in optical lattices [4], which allow the time resolved observation of coherent quantum dynamics [5].

The motivation of the work in this thesis is partly based on the idea that the quantum mechanical state of a MBS may become useful for novel protocols that can be loosely grouped under the term *quantum communication*. Two suitable degrees of freedom of a single subunit of the MBS, like the $|\uparrow\rangle$ and $|\downarrow\rangle$ states of an atom with effective spin-1/2, as well as two of its electronic states, can form a *qubit*—the elementary unit of quantum information which can be parameterised by two complex amplitudes

$$|\phi\rangle = \alpha|\uparrow\rangle + \beta|\downarrow\rangle. \tag{1.1}$$

For the purpose of carrying out quantum communication tasks, of which we will briefly outline a prominent example below, spatially separated qubits are usually required to become maximally entangled. The Bell states

$$|\psi^{\pm}\rangle = \frac{1}{\sqrt{2}}(|\uparrow\downarrow\rangle \pm |\downarrow\uparrow\rangle) \tag{1.2}$$

H.C. Wichterich, *Entanglement Between Noncomplementary Parts of Many-Body Systems*, Springer Theses, DOI: 10.1007/978-3-642-19342-2_1,
© Springer-Verlag Berlin Heidelberg 2011

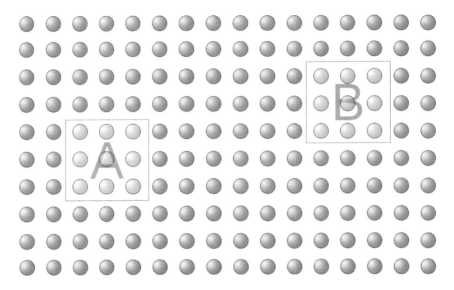

Fig. 1.1 Schematic of a rectangular lattice of interacting particles. Typical quantum mechanical states of this type of many-body system, including its one and three dimensional analogues, contain entanglement [1], a form of correlation by which measurable quantities of spatially distinct regions A and B, say, become intertwined with each other in a profound way [6, 7]. At the interface of many-particle physics [8] and quantum information science [9] entanglement acts as a sort of probe of non-local correlations in the many-body wave-function and its behaviour may signal fundamental changes in the physical behaviour of strongly correlated systems. For instance, entanglement is regarded to be playing a mayor role in physical phenomena such as exotic states of matter (fractional quantum Hall states, frustrated magnets) and also in phase transitions at zero temperature, realising a class of markedly non-classical states of matter

$$|\phi^{\pm}\rangle = \frac{1}{\sqrt{2}}(|\uparrow\uparrow\rangle \pm |\downarrow\downarrow\rangle) \qquad (1.3)$$

constitute a basis of such maximally entangled states of two qubits. The state $|\psi^-\rangle$ is also called *singlet* while the remaining three are grouped under the name *triplet*. A consequence of entanglement is that the individual state of each member of a Bell pair is maximally uncertain, in that there is a 50% chance of measuring $|\uparrow\rangle$ or $|\downarrow\rangle$ on an individual qubit of a Bell pair (pair of qubit particles in one of the Bell states).

One example for a typical objective of quantum communication is that of transferring a general qubit state $|\phi\rangle$ (1.1) from a sender (Alice) to a receiver (Bob) as accurately as possible. To accomplish this, Alice can simply encode $|\phi\rangle$ on a carrier, e.g. an atom or a photon, and send it down a channel. Alternatively, Alice and Bob can use teleportation [10] which is a means to transfer state $|\phi\rangle$ from one point in space to another by only classical communication and a quantum resource. We will briefly review this protocol hereafter, along the lines of the original work [10]. Prior to teleportation, the two parties need to share a pair of particles in a maximally entangled state, for example the Bell singlet state $|\psi^-\rangle$ of

(1.2). So initially the state of the three involved particles, that is Alice's particle 1 which is initially the carrier of quantum state $|\phi\rangle$ and the entangled pair of particles labelled 2 and 3, is given by

$$|\varphi\rangle_{123} = |\phi\rangle_1 \otimes |\psi^-\rangle_{23} = \frac{\alpha}{\sqrt{2}}(|\uparrow_1\uparrow_2\downarrow_3\rangle - |\uparrow_1\downarrow_2\uparrow_3\rangle) - \frac{\beta}{\sqrt{2}}(|\downarrow_1\uparrow_2\downarrow_3\rangle - |\downarrow_1\downarrow_2\uparrow_3\rangle)$$

where the subscript indices label the considered particle. In terms of $|\psi^\pm\rangle_{12}$ and $|\phi^\pm\rangle_{12}$ the three particle state reads as

$$|\varphi\rangle_{123} = \frac{1}{2}\left[|\psi^-\rangle_{12}|\phi^{(a)}\rangle_3 + |\psi^+\rangle_{12}|\phi^{(b)}\rangle_3 + |\phi^-\rangle_{12}|\phi^{(c)}\rangle_3 + |\phi^+\rangle_{12}|\phi^{(d)}\rangle_3\right]$$

where

$$|\phi^{(a)}\rangle_3 = -\alpha|\uparrow\rangle_3 - \beta|\downarrow\rangle_3$$
$$|\phi^{(b)}\rangle_3 = -\alpha|\uparrow\rangle_3 + \beta|\downarrow\rangle_3$$
$$|\phi^{(c)}\rangle_3 = \alpha|\uparrow\rangle_3 + \beta|\downarrow\rangle_3$$
$$|\phi^{(d)}\rangle_3 = \alpha|\uparrow\rangle_3 - \beta|\downarrow\rangle_3.$$

If Alice were to perform a Bell-type measurement on particles 1 and 2, thereby preparing them in one of the states $\{|\psi^\pm\rangle_{12}, |\phi^\pm\rangle_{12}\}$, the state of particle 3 would be determined by her measurement outcome ($x = a, b, c,$ or d) by virtue of the von Neumann measurement postulate [12].

The measurement outcome could be encoded in two classical bits, and would enable Bob to produce a replica of $|\phi\rangle$ from the post-measurement state $|\phi^{(x)}\rangle_3$ of particle 3. Each $|\phi^{(x)}\rangle_3$ is related to the original state by specific qubit rotations.

Hence, $|\psi^-\rangle$ enables the noiseless transmission of a state $|\phi\rangle$ from Alice to Bob Fig. (1.2). The capacity of the quantum resource, that is $|\psi^-\rangle$, is underlined by realising that from knowing only two classical bits (which suffice to encode the measurement outcome, see Fig. 1.2) the receiver can reconstruct a state that is described by two real parameters.[1]

A pressing question is whether the entanglement appearing naturally in systems of interacting particles can be exploited for genuinely nonclassical communication protocols such as quantum teleportation. Firstly, this goal is hampered by the fact that the state of each pair of particles will be a *mixture*, and statistical uncertainty regarding the particular microscopic state of the pair degrades the hypothetically achievable entanglement of pure Bell states of (1.2) and (1.3). It is therefore important to be provided with a criterion of whether a mixed quantum state still contains sufficient entanglement in the sense that it may be useful in a quantum communication context. One such criterion is that of identifying a state as being

---

[1] These are e.g. azimuth and polar angle in the Bloch sphere representation of the qubit state. Clearly, in order for the receiver to learn about these values, repeated teleportation experiments would have to be carried out.

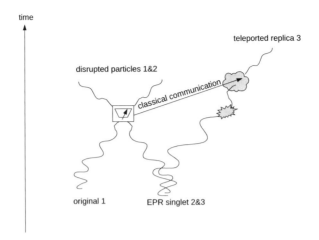

Fig. 1.2 Quantum teleportation. A pair of particles (2 and 3) is prepared in a Bell singlet state $|\psi^-\rangle$ of (1.2). One member of this pair, labelled 2 is given to Alice (*left*) and the other—labelled 3—is given to Bob (*right*). Alice possesses a second particle, designated 1, the qubit state $|\phi\rangle$ of which she aims to transfer to Bob. To this end, she performs a Bell-type measurement on the two particles 1 and 2 so as to prepare them in one of the four Bell states $|\psi^\pm\rangle, |\phi^\pm\rangle$. This disentangles particle 3 from 2. Alice's measurement outcome is reported to Bob by way of classical communication, enabling him to reconstruct the state $|\phi\rangle$ by means of single qubit rotations on the post-measurement state of particle 3 (for details, see [10]). Figure reproduced from [11]

purifiable: Given a number of copies of a state of two qubits this state is said to be purifiable if one can extract from them a lesser number of almost pure Bell states. For this to be possible, certain conditions must be met as will be discussed briefly in Sect. 2.5.

In the literature, several ways of creating entanglement between particles at distances larger than a few lattice spacings were proposed. An example for a low control[2] means of entangling remote sites in a spin chain is to distribute entanglement dynamically, e.g. in terms of elementary excitations (magnons, which travel wave-like across the lattice). This was first proposed in the context of a quantum state transfer protocol [13]. In a similar way, we rely on an out-of-equilibrium situation in Chap. 2 of the present work by investigating the effects of a so-called *quantum quench* (i.e. a rapid change in interaction parameters, thereby usually crossing a border which separates different magnetic phases).

Another prominent approach of entangling two particular distant spins in a collection of interacting spins is based on the concept of localisable entanglement [14, 15]. This entanglement is achieved by way of local measurements on all but the two designated spins, though the task of addressing individual sites of a bulk system can be challenging. One can also ask whether a measurement can be instrumental if it is performed on regions A and B themselves, comprising several

---

[2] Referring to scenarios where access and control of single subunits inside the bulk of the many-body system is not strictly required, an exception being, for instance, the terminal sites.

spins each. In this work, we study this latter question in a particular model and we see that measurement can lead to a pure form of entanglement despite the initial (unperturbed) state of the regions is mixed.

Entanglement between the individual constituent particles of a MBS at its zero temperature equilibrium state is usually very short ranged, often extending no further than over a distance of a few inter-particle spacings [16, 17]. There are exceptions to this rule, though: Long-range entanglement can be induced between the spins at the terminal sites of an open ended chain of spin-1/2 particles by engineering rather artificial, i.e. strongly non-uniform, couplings [18].

Alternatively, one can study the entanglement shared between large groups of particles rather than individual pairs. Here, we look at large regions of spins of uniformly coupled spin chains and show that, even though no pair of single spins between the regions is entangled in the ground state, as a collection they do share entanglement, especially when the system undergoes a quantum phase transition. The entanglement of macroscopic groups of spins at quantum phase transitions displays interesting properties which we study in a highly simplified, but analytically tractable model of strongly correlated spins.

Summary of results. The study on quantum quench presented in Chap. 2 shows, that under idealised conditions purifiable entanglement can be dynamically established between the remotest pair of spins in XX spin chains of up to $N = 241$ spins. For shorter chains of up to ~ 13 spins displays a fully entangled fraction of $f \gtrsim 0.7$ and the edge spins assume a state which is well approximated by $\rho_{1N} = f|\psi^+\rangle\langle\psi^+| + (1-f)/2(|\phi^+\rangle\langle\phi^+| + |\phi^-\rangle\langle\phi^-|)$. We evidence that the underlying mechanisms are commensurate with a picture of entangled quasi-particles travelling with constant velocity and opposite orientation across the lattice, giving rise to the observed high amount of entanglement between the terminal sites of the spin chain at an optimal time $T_{\max}$ which scales linearly with N. It is shown that this entanglement is very robust towards detrimental effects like random disorder and could be evidenced in various experimental setups, including cold atoms, trapped ions or superconducting circuits.

The extraction of pure state entanglement from distinguished regions of MBS by way of ideal quantum measurements is the subject of Chap. 3. We show that for super singlet states of three qutrits as well as the ground state of the transverse XY spin ring model this extraction scheme is probabilistically feasible. Hence, local measurements in typical many-body states can lead to pure state entanglement between the degrees of freedom of possibly well separated regions of MBS, given that prior to the measurement the quantum state of the regions carries some form of (noisy) entanglement. This latter requirement constitutes a major difference to localisable entanglement which can potentially establish entanglement between regions which were in a separable state before the protocol.

Chapters 4 and 5 are dealing with universal scaling properties of the negativity, a measure which we invoke in order to quantify the degree of mixed state entanglement of disjoint (or noncomplementary) regions of models of spin-1/2 particles undergoing a quantum phase transition. Based on numerical evidence provided in

Chap. 4 we conjecture that negativity between large noncomplementary regions of transverse XY spin chains depends only on the relative sizes of the involved regions and is, hence, manifestly independent of the overall system size if these ratios are fixed. We make an ansatz for the decay of negativity as a function of the separation of the entangled regions and their respective extent, which can be accurately fitted to the numerical data. Strikingly, even at criticality this form of correlation decays exponentially with distance, a rather disturbing finding considering that classical correlation functions decay as a power law at the transition. Also, at the quantum critical points negativity is found to be largely independent of the microscopic details of the model which is a signature of universality. In Chap. 5 this universality is verified in a collective spin model, the Lipkin–Meshkov–Glick model, which bears strong similarities to the XY spin chain except for the long range interactions. Here, the thermodynamic limit can be taken conveniently and an analytic solution of negativity is obtained across the entire phase diagram. It is further remarkable that negativity of noncomplementary regions is manifestly finite at the quantum phase transitions and in the thermodynamic limit, substantiating our conjecture that in spin chain model of Chap. 4 negativity remains finite as well.

Outline. After a brief introduction on quantification of entanglement (Sect. 1.1) we present our results on long-range entanglement established dynamically after quantum quench (Chap. 2). We then discuss entanglement of noncomplementary parts of a spin chain in the context of distant von Neumann measurements (Chap. 3), and finally study its role at continuous quantum phase transitions both in a spin chain model with nearest neighbour interactions (Chap. 4) and the corresponding model with infinite range interactions (Chap. 5).

The work in this thesis has been done under supervision and in collaboration with Prof. Sougato Bose. Contributions from other collaborators will be indicated in the individual chapters where appropriate.

1.1 Quantifying Entanglement in Many-Body Systems

The subject of quantifying entanglement in generic states of MBS is an extremely demanding task. In this section, we introduce the entanglement measures that will be extensively used throughout this work. We will discuss their properties and relate them to other measures that can be found in the literature.

1.1.1 Separability

A generic state will be a statistical mixture represented by a density operator[3]

---

[3] Here and in the following " ≥ 0 " for operators (and scalars) means non-negative or positive semidefinite.

$$\hat{\rho} \geq 0, \quad \hat{\rho} = \hat{\rho}^\dagger \tag{1.4}$$

which obeys

$$\mathrm{Tr}\big[\hat{\rho}^2\big] \leq \mathrm{Tr}[\hat{\rho}] = 1. \tag{1.5}$$

Equality in (1.5) holds if and only if the state is *pure* [19]. Let us define what an entangled state is. This is conventionally done by defining what it is not, namely separable. So [20].

Definition 1.1.1 *(Separability)* A state $\hat{\rho} \in \mathcal{H}_A \otimes \mathcal{H}_B$ is called separable if and only if it can be written as

$$\hat{\rho} = \sum_k p_k \, \hat{\rho}_A^{(k)} \otimes \hat{\rho}_B^{(k)}, \tag{1.6}$$

$$\hat{\rho}_A^{(k)} \in \mathcal{H}_A, \ \hat{\rho}_B^{(k)} \in \mathcal{H}_B$$

with probabilities $p_k \in [0,1]$, $\sum_k p_k = 1$ and pure density matrices $\hat{\rho}_A^{(k)}$ and $\hat{\rho}_B^{(k)}$. Two distant observers could prepare a separable state by following instructions from a third party on how to prepare—locally—their respective subsystem. If the state is not separable the state is called entangled.

1.1.2 Entropy of Entanglement

Pure states of composite quantum systems constitute an exception regarding the relative ease with which we can quantify entanglement between constituents (say, that between a distinguished region, the system S, and its complement, the environment E): The more these two parties are entangled, the more mixed are the states of the subsystems S and E when considered individually.

A pure state on a bipartite Hilbert space

$$|\psi\rangle \in \mathcal{H}_S \otimes \mathcal{H}_E$$

of dimension $\dim \mathcal{H} = d = d_S d_E$ can be decomposed as

$$|\psi\rangle = \sum_{k,l} \psi_{k,l}|k\rangle \otimes |l\rangle \tag{1.7}$$

where $|k\rangle$ and $|l\rangle$ denote basis vectors of $\mathcal{H}_S$ and $\mathcal{H}_E$, respectively. This expression can be simplified by invoking the singular value decomposition (SVD) [21] of the matrix of coefficients

$$\psi_{k,l} = (UDV^\dagger)_{k,l}$$

where $D = \mathrm{diag}(\sqrt{w_1}, \sqrt{w_2}, \ldots, \sqrt{w_\chi})$ is a positive, diagonal matrix of dimension $\chi \times \chi$, $\chi \leq \min(d_S, d_E)$ and U and V are unitary matrices.[4] Forming linear combinations of the $|k\rangle$ in (1.7) with the columns of U and similarly the $|l\rangle$ with the rows of $V^\dagger$ leads to the Schmidt decomposition (SD)

$$|\psi\rangle = \sum_{\alpha=1}^{\chi} \sqrt{w_\alpha} |w_\alpha^S\rangle \otimes |w_\alpha^E\rangle \tag{1.8}$$

$$|w_\alpha^S\rangle = \sum_k U_{k,\alpha} |k\rangle \tag{1.9}$$

$$|w_\alpha^E\rangle = \sum_l V_{l,\alpha}^* |l\rangle. \tag{1.10}$$

We will call the $\sqrt{w_\alpha}$ the Schmidt values, χ the Schmidt rank, and $|w_\alpha^S\rangle, |w_\alpha^E\rangle$ the Schmidt vectors. It follows from the separability criterion 1.1.1 that $|\psi\rangle$ is separable if and only if the Schmidt rank is one [20].

Entropy of entanglement quantifies the entanglement between S and E in such a bipartite pure state $\hat{\rho} = |\psi\rangle\langle\psi|$ [20]

$$\mathcal{E}(\hat{\rho}) = -\mathrm{Tr}[\hat{\rho}_S \ln \hat{\rho}_S], \quad \hat{\rho}_S = \mathrm{Tr}_E[\hat{\rho}] = \sum_{\alpha=1}^{\chi} w_\alpha |w_\alpha^S\rangle\langle w_\alpha^S|. \tag{1.11}$$

The operator $\hat{\rho}_S$ designates the reduced density operator describing the state of the system S and the operation $\mathrm{Tr}_E[\cdots]$ is called *partial trace* and amounts to a trace with respect to the environmental degrees of freedom only. Equation 1.11 highlights two important properties of the SD.

- The reduced density operator $\hat{\rho}_S$ is diagonal in the basis of Schmidt vectors $|w_\alpha^S\rangle$.
- The eigenvalues of $\hat{\rho}_S$ are given by the squares of the Schmidt values $\sqrt{w_\alpha}$.

Analogous conclusions follow for the reduced density operator of the environment. Hence, in terms of the eigenvalues w_α of $\hat{\rho}_S$ (and $\hat{\rho}_E$) entropy of entanglement assumes

$$\mathcal{E}(\hat{\rho}) = -\sum_\alpha w_\alpha \ln w_\alpha. \tag{1.12}$$

Alternatively, it suffices to know $\mathrm{Tr}[\hat{\rho}_S^n]$ so that entropy of entanglement follows from

$$\mathcal{E}(\hat{\rho}) = -\partial_n \mathrm{Tr}[\hat{\rho}_S^n]|_{n=1}, \tag{1.13}$$

(replica trick, see [3]).

---

[4] This variant of the SVD is sometimes called compact or economic SVD.

1.1.3 Entanglement of Formation and Entanglement Cost

It is now important to realise that if the state $\hat{\rho}$ of the global system $(S + E)$ is a statistical mixture, then $\mathcal{E}$ is no longer a meaningful measure of the entanglement between S and E. For example, let S and E be a two-level system (or qubit) each, whose basis states are $|\uparrow\rangle$ and $|\downarrow\rangle$, respectively. An equal mixture of $|\uparrow\uparrow\rangle$ and $|\downarrow\downarrow\rangle$ leads to maximally mixed states of either party. Yet their state is fully separable.

Entanglement of formation (EOF), designated by the symbol $\mathcal{E}_F$, is a fundamental measure of entanglement in mixed quantum states $\hat{\rho}$ and is defined as the least expected entanglement within all possible ensembles of pure states which realise $\hat{\rho}$

$$\mathcal{E}_F(\hat{\rho}) = \min\left\{ \sum_i p_i \mathcal{E}(|\psi_i\rangle\langle\psi_i|) \;\Big|\; \hat{\rho} = \sum_i p_i |\psi_i\rangle\langle\psi_i| \right\}. \qquad (1.14)$$

Explicit evaluation of $\mathcal{E}_F$ has been successful only in very few cases, for instance for systems of two qubits [22]. An important operational interpretation of EOF was achieved in [23] by proving the following identity of entanglement cost

$$\mathcal{E}_C(\hat{\rho}) = \lim_{n\to\infty} \frac{\mathcal{E}_F(\hat{\rho}^{\otimes n})}{n}, \qquad (1.15)$$

where $\hat{X}^{\otimes n}$ denotes the n-fold tensor product of $\hat{X}$ with itself. In simple terms, $\mathcal{E}_C$ quantifies the least number of initial qubits that need to be communicated to prepare a state $\hat{\rho}$. By virtue of teleportation, the number of communicated qubits is equivalent to the number of initial Bell pairs that need to be shared between the parties. For a detailed discussion on entanglement cost, we refer to [20, 24].

1.1.4 Separability and Positive Maps

A powerful result to decide the separability of general states on bipartite Hilbert spaces is the following [25]

Theorem 1.1.2 (Horodecki) *A state $\hat{\rho}$ acting on a Hilbert space $\mathcal{H} = \mathcal{H}_A \otimes \mathcal{H}_B$ is separable if and only if*

$$(\Phi \otimes \mathbb{1}_B)\hat{\rho} \geq 0$$

for all positive maps Φ acting on operators on $\mathcal{H}_A$. $\mathbb{1}_B$ is the identity map on $\mathcal{H}_B$.

We say a map is positive if it maps positive operators into positive operators, i.e. $\Phi(\hat{X}) \geq 0 \;\forall\; \hat{X} \geq 0$. Sampling all positive maps is rarely viable, but if for a particular positive map Φ the operator $(\Phi \otimes \mathbb{1}_B)\hat{\rho}$ has negative eigenvalues (under these circumstances Φ is positive but not *completely positive*), one has shown that

Fig. 1.3 Quantum
mechanical states can be
categorised as being
separable or entangled states.
PPT states are states with
positive partial transpose and
need not be separable. PPT
states that are entangled are
called PPT entangled states

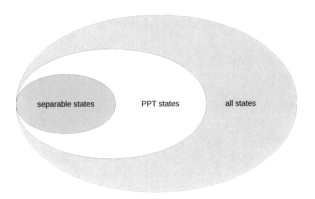

$\hat{\rho}$ is surely entangled. One such example for a positive map is the transposition
map T, and we will use the following notation in the following

$$\hat{\rho}^{T_A} \equiv (T \otimes \mathbb{1}_B)\hat{\rho} \tag{1.16}$$

and call this map *partial transposition*. In a product basis spanned by vectors
$|k\rangle \otimes |l\rangle \equiv |k, l\rangle$ on the bipartite Hilbertspace $\mathcal{H}_{AB} = \mathcal{H}_A \otimes \mathcal{H}_B$, partial transposition of a general state $\hat{\rho}$ would amount to

$$\langle k, l|\hat{\rho}^{T_A}|m, n\rangle = \langle m, l|\hat{\rho}|k, n\rangle.$$

The separability criterion $\hat{\rho}^{T_A} \geq 0$ is known as the positive partial transpose
(PPT) criterion [25, 26] and is a necessary but not sufficient criterion, in general.
The class of states which satisfy $\hat{\rho}^{T_A} \geq 0$ but are not separable are called PPT-
entangled states Fig. 1.3.

In summary,

$$\hat{\rho} \text{ separable } \Rightarrow \hat{\rho}^{T_A} \geq 0 \tag{1.17}$$

$$\hat{\rho} \text{ entangled } \Leftarrow \hat{\rho}^{T_A} \text{ has negative eigenvalues.} \tag{1.18}$$

1.1.5 Negativity and Logarithmic Negativity

The results on positive maps presented in the previous section can also be put to
use for the quantification of entanglement in mixed states. The degree of violation
of positivity, as measured by the trace norm

$$\|\hat{\rho}^{T_A}\| = \text{Tr}\left[\sqrt{\left(\hat{\rho}^{T_A}\right)^\dagger \hat{\rho}^{T_A}}\right] = \text{Tr}\left[\sqrt{\left(\hat{\rho}^{T_A}\right)^2}\right] \tag{1.19}$$

is immediately related to the negativity [27–29]

$$\mathcal{N}(\hat{\rho}) \equiv \|\hat{\rho}^{T_A}\| - 1 \tag{1.20}$$

and its close relative, the logarithmic negativity [29, 30]

$$\mathcal{L}(\hat{\rho}) \equiv \ln \|\hat{\rho}^{T_A}\|. \tag{1.21}$$

Partial transposition preserves the trace $\mathrm{Tr}[\hat{\rho}^{T_A}] = \mathrm{Tr}[\hat{\rho}] = 1$, so that in terms of eigenvalues v_α of $\hat{\rho}^{T_A}$ negativity assumes

$$\mathcal{N}(\hat{\rho}) = 2 \sum_{v_\alpha < 0} |v_\alpha|, \tag{1.22}$$

hence negativity is given by two times the sum of moduli of negative eigenvalues of $\hat{\rho}^{T_A}$ which justifies the nomenclature.

Negativity attains its maximum for pure states [29]. In terms of the SD (1.8) partial transposition on subsystem E amounts to

$$\hat{\rho}^{T_E} = (|\psi\rangle\langle\psi|)^{T_E} = \sum_{\alpha,\alpha'} \sqrt{w_\alpha w_{\alpha'}} \, |w_{\alpha'}^S, w_\alpha^E\rangle\langle w_\alpha^S, w_{\alpha'}^E|. \tag{1.23}$$

Eigenvectors and eigenvalues of $\hat{\rho}^{T_E}$ are [31]

$$|w_\alpha^S, w_\alpha^E\rangle \quad \text{eigenvalue} \quad w_\alpha \tag{1.24}$$

$$\frac{1}{\sqrt{2}} \left(|w_{\alpha'}^S, w_\alpha^E\rangle \pm |w_\alpha^S, w_{\alpha'}^E\rangle \right) \quad \text{eigenvalue} \pm \sqrt{w_\alpha w_{\alpha'}} \; (\alpha < \alpha') \tag{1.25}$$

giving rise to a total of $\chi + 2\frac{(\chi-1)\chi}{2} = \chi^2$ eigenvalues which exhausts the maximal number of nonzero eigenvalues.

This leads to the negativity of bipartite pure state of (1.8)

$$\mathcal{N}(|\psi\rangle\langle\psi|) = 2 \sum_{\alpha < \alpha'} \sqrt{w_\alpha w_{\alpha'}} = \left(\sum_{\alpha,\alpha'} \sqrt{w_\alpha w_{\alpha'}} \right) - 1 = \left(\sum_\alpha \sqrt{w_\alpha} \right)^2 - 1 \tag{1.26}$$

where we summed over the moduli of negative eigenvalues (compare (1.22)). Owing to the concavity of the square root, one has that

$$\sum_{\alpha=1}^{\chi} \sqrt{w_\alpha} \leq \sqrt{\chi}$$

where the equality holds if all eigenvalues $w_\alpha = 1/\chi$, corresponding to a maximally entangled state. This implies the following relation for negativity

$$0 \leq \mathcal{N}(\hat{\rho}) \leq \chi - 1, \tag{1.27}$$

where zero is assumed for separable pure states having Schmidt rank $\chi = 1$. Accordingly, logarithmic negativity of a pure state reads as [29]

$$\mathcal{L}(|\psi\rangle\langle\psi|) = 2\ln\left(\sum_\alpha \sqrt{w_\alpha}\right) \tag{1.28}$$

and provides an upper bound on entropy of entanglement [29]

$$\mathcal{E}(|\psi\rangle\langle\psi|) = 2\sum_\alpha w_\alpha \ln\left(\frac{1}{\sqrt{w_\alpha}}\right) \tag{1.29}$$

$$\leq 2\ln\left(\sum_\alpha \sqrt{w_\alpha}\right) = \mathcal{L}(|\psi\rangle\langle\psi|). \tag{1.30}$$

For pure states, both $\mathcal{N}$ and $\mathcal{L}$ can be expressed in terms of Rényi entropy

$$\mathcal{E}^{(n)}(\hat{\rho}) \equiv \frac{1}{1-n}\ln \mathrm{Tr}[\hat{\rho}_S^n] \tag{1.31}$$

$$\Leftrightarrow \quad \mathcal{L} = \mathcal{E}^{(1/2)}(\hat{\rho}), \quad \mathcal{N} = \exp\left(\mathcal{E}^{(1/2)}(\hat{\rho})\right) - 1 \tag{1.32}$$

hence we realise that, in general, neither negativity nor logarithmic negativity reduces to the entropy of entanglement ($\mathcal{E} = \lim_{n\to 1}\mathcal{E}^{(n)}$) which is sometimes considered a deficiency. Only in the limit of maximally entangled states with $w_\alpha = 1/\chi$, logarithmic negativity equals the entropy of entanglement.

1.1.6 Entanglement Monotones, Entanglement Measures and Bounds

We discuss here several requirements that are commonly imposed on measures of entanglement, i.e. magnitudes which ideally quantify how useful a shared state of two parties is in terms of a communication resource. For a potential measure $E(\hat{\rho})$ to qualify as *entanglement monotone*, the following two axioms were suggested to be mandatory [20]:

- *Monotonicity under LOCC.* $E(\hat{\rho})$ does not increase, on average, under local operations and classical communication.
- $E(\hat{\rho})$ *vanishes on separable states.*

On these grounds, both negativity and logarithmic negativity are entanglement monotones [28, 30]. In addition to the above two axioms, considerably stricter requirements on $E(\hat{\rho})$ are imposed for it to be called an *entanglement measure* [32, 33]:

- *Discriminance.* $E(\hat{\rho}) = 0$ if and only if ρ is separable.
- *Convexity.* $E(a\hat{\rho} + (1-a)\hat{\sigma}) \leq aE(\hat{\rho}) + (1-a)E(\hat{\sigma})$ for $a \in [0,1]$.
- Additivity. $E(\hat{\rho} \otimes \hat{\rho}) = 2E(\hat{\rho})$.

- $E(\hat{\rho})$ reduces to entanglement entropy for all pure states ρ

To the present day, the research community has not reached a consensus as to which axioms are to be considered mandatory, and consequently whether negativity and logarithmic negativity are to be regarded as entanglement measures, in that, for example, negativity is not additive and logarithmic negativity is not convex and neither of them can fully discriminate the set of separable states from the set of entangled states. This is because, by construction, they can not detect PPT entangled states. Further, as we have seen in the foregoing section, neither of them reduces to the entanglement entropy on pure states, in general. Often the ease of their computation (both numerically and analytically) has vindicated their widespread use in the literature (see list of citing articles of [29]) and has out-weighed (from practitioners' perspective) the mentioned concerns.

An important vindication for negativity and logarithmic negativity, above their practicality, is that lower bounds on EOF can be constructed from $\|\hat{\rho}^{T_A}\|$ [34]. Since the EOF can often not be computed explicitly, finding tight bounds on meaningful measures is a matter of great importance. In conclusion, we adopt a utilitarian viewpoint with respect to negativity and logarithmic negativity and disregard certain deficiencies when it comes to axiomatic rigour and operational interpretation. Our justification of using them is their capacity of providing bounds to more widely accepted monotones.

References

1. L. Amico, R. Fazio, A. Osterloh, V. Vedral, Entanglement in many-body systems. Rev. Mod. Phys. **80**, 517 (2009)
2. J. Eisert, M. Cramer, M.B. Plenio, Colloquium: area laws for the entanglement entropy. Rev. Mod. Phys. **82**(1), 277–306 (2010)
3. J. Cardy, Entanglement entropy in extended quantum systems. Eur. Phys. J. B. **64**, 321–326 (2008)
4. I. Bloch, Quantum gases. Science **319**(5867), 1202–1203 (2008)
5. S. Trotzky, P. Cheinet, S. Folling, M. Feld, U. Schnorrberger, A.M. Rey, A. Polkovnikov, E.A. Demler, M.D. Lukin, I. Bloch, Time-resolved observation and control of superexchange interactions with ultracold atoms in optical lattices. Science **319**(5861), 295–299 (2008)
6. A. Einstein, B. Podolsky, N. Rosen, Can quantum-mechanical description of physical reality be considered complete? Phys. Rev. **47**(10), 777–780 (1935)
7. J.S. Bell, On the Einstein–Podolsky–Rosen paradox. Physics **1**, 195–200 (1964)
8. J.W. Negele, H. Orland, *Quantum Many-Particle systems*. (Addison Wesley, Reading, MA, 1988)
9. M.A. Nielsen, I.L. Chuang, *Quantum Computation and Quantum Information*. (Cambridge University Press, Cambridge, 2000)
10. C.H. Bennett, G. Brassard, C. Crépeau, R. Jozsa, A. Peres, W.K. Wootters, Teleporting an unknown quantum state via dual classical and Einstein-Podolsky-Rosen channels. Phys. Rev. Lett. **70**(13), 1895–1899 (1993)
11. IBM Research. http://www.research.ibm.com/quantuminfo/teleportation/
12. C. Cohen-Tannoudji, B. Diu, F. Laloe, *Quantum Mechanics*. Wiley-Interscience, New York, 2006).

13. S. Bose, Quantum communication through an unmodulated spin chain. Phys. Rev. Lett. **91**(20), 207901 (2003)
14. F. Verstraete, M. Popp, J.I. Cirac, Entanglement versus correlations in spin systems. Phys. Rev. Lett. **92**(2), 027901 (2004)
15. M. Popp, F. Verstraete, M.A. Martín-Delgado, J.I. Cirac, Localizable entanglement. Phys. Rev. A **71**(4), 042306 (2005)
16. T.J. Osborne, M.A. Nielsen, Entanglement in a simple quantum phase transition. Phys. Rev. A **66**(3), 032110 (2002)
17. A. Osterloh, L. Amico, G. Falci, R. Fazio, Scaling of entanglement close to a quantum phase transition. Nature **416**, 608–610 (2002)
18. L. Campos Venuti, C. DegliEsposti Boschi, M. Roncaglia, Long-distance entanglement in spin systems. Phys. Rev. Lett. **96**(24), 247206 (2006)
19. L.E. Ballentine, *Quantum Mechanics: A modern development*. (World Scientific Publishing Co. Pte. Ltd, Singapore, 1998)
20. R. Horodecki, P. Horodecki, M. Horodecki, K. Horodecki, Quantum entanglement. Rev. Mod. Phys. **81**, 865 (2009)
21. G.H. Golub, van C.F. Loan, Matrix Computations, 3rd edn. (John Hopkins University Press, Baltimore, 1996)
22. W.K. Wootters, Entanglement of formation of an arbitrary state of two qubits. Phys. Rev. Lett. **80**(10), 2245–2248 (1998)
23. P.M. Hayden, M. Horodecki, B.M. Terhal, The asymptotic entanglement cost of preparing a quantum state. J. Phys. A: Math. Gen. **34**(35), 6891 (2001)
24. D. Bruss, G. Leuchs, *Lectures on Quantum Information*. (Wiley-VCH, Weinheim, 2007)
25. M. Horodecki, P. Horodecki, R. Horodecki, Separability of mixed states: necessary and sufficient conditions. Phys. Lett. A **223**(1–2), 1–8 (1996)
26. A. Peres, Separability criterion for density matrices. Phys. Rev. Lett. **77**(8), 1413–1415 (1996)
27. K. Życzkowski, P. Horodecki, A. Sanpera, M. Lewenstein, Volume of the set of separable states. Phys. Rev. A **58**(2), 883–892 (1998)
28. J. Lee, M.S. Kim, H. Jeong, Partial teleportation of entanglement in the noisy environment. J. Mod. Opt. **47**, 2151 (2000)
29. G. Vidal, R.F. Werner, Computable measure of entanglement. Phys. Rev. A **65**(3), 032314 (2002)
30. M.B. Plenio, Logarithmic negativity: a full entanglement monotone that is not convex. Phys. Rev. Lett. **95**(9), 090503 (2005)
31. A. Datta, *Studies on the Role of Entanglement in Mixed-State Quantum Computation*. (The University of New Mexico, Albuquerque, 2008)
32. I. Bengtsson, K. Zyczkowski, *Geometry of Quantum States: an Introduction to Quantum Entanglement*. (Cambridge University Press, Cambridge, 2006)
33. G. Vidal, Entanglement monotones. J. Mod. Opt. **47**(2/3), 355 (2000)
34. K. Chen, S. Albeverio, S. M Fei, Entanglement of formation of bipartite quantum states. Phys. Rev. Lett. **95**(21), 210501 (2005)

Chapter 2
Exploiting Quench Dynamics in Spin Chains for Distant Entanglement and Quantum Communication

In this chapter we propose a method to entangle two distant spin-1/2 particles in a one dimensional array of spins through *quantum quench* [1–5]. A quench is the rapid change of internal or external parameters which govern the physics of a system of interacting particles. This change of parameters could concern, for instance, an alteration of competing particle–particle interactions or the change of the amplitude of external fields, the temperature, pressure, and so forth. When a quench is performed at zero temperature one refers to a quantum quench. In this idealised case, the system would initially reside in its ground state and would subsequently display a *coherent* response due to quench. A strong response is expected if the initial and final set of parameters which define the quench correspond to distinct phases that the interacting system would equilibrate towards. An example is the superfluid state and the Mott insulator state of a Bose Einstein condensate where a transition between them is triggered by changing the depth of an optical lattice potential [6]. The dynamics which occur in the course of such phase transitions have been of great interest in the context of defect formation and a variety of physical systems [7].

The study of quantum quenches was recently fuelled by the prospect of realising them in experiments on cold atoms which are trapped in the periodic troughs of a potential landscape created by standing electromagnetic waves [6, 8]. These experiments allow the observation of coherent quantum evolution of strongly correlated atomic systems in conjunction with a high degree of experimental control [9]. Magnetic exchange interactions among the constituent particles where evidenced in [10] and are dynamically tunable [11–13]. Hence, these systems are an almost ideal testbed for theoretical predictions of coherent effects after a quantum quench. We briefly review some alternative experimental setups towards the end of this chapter, in Sect. 2.6.3.

Following a brief discussion of the model under consideration (Sect. 2.1) our scheme to entangle distant spins exploiting a quantum quench will be explained in Sect. 2.2 Some methodical detail will be presented in Sects. 2.3 and 2.4. We

H.C. Wichterich, *Entanglement Between Noncomplementary Parts of Many-Body Systems*, Springer Theses, DOI: 10.1007/978-3-642-19342-2_2,
© Springer-Verlag Berlin Heidelberg 2011

introduce the notion of purifiable entanglement in Sect. 2.5. The chapter closes
with the presentation and discussion of our results in Sect. 2.6.

2.1 The XXZ Spin Chain Model

The model under consideration is an open ended chain of N spin-$\frac{1}{2}$ systems with
nearest-neighbour XXZ interaction

$$\hat{H} = \sum_{l=1}^{N-1} \frac{J}{2} \left(\hat{\sigma}_l^x \hat{\sigma}_{l+1}^x + \hat{\sigma}_l^y \hat{\sigma}_{l+1}^y + \Delta \hat{\sigma}_l^z \hat{\sigma}_{l+1}^z \right) \tag{2.1}$$

where the parameters J and Δ denote the coupling strength and the anisotropy,
respectively. The Pauli operators are defined by their action on the computational
basis states $\{|\uparrow\rangle, |\downarrow\rangle\}$ of a spin at site l

$$\hat{\sigma}_l^x |\uparrow\rangle = |\downarrow\rangle, \quad \hat{\sigma}_l^x |\downarrow\rangle = |\uparrow\rangle \tag{2.2}$$

$$\hat{\sigma}_l^y |\uparrow\rangle = i|\downarrow\rangle, \quad \hat{\sigma}_l^y |\downarrow\rangle = -i|\uparrow\rangle \tag{2.3}$$

$$\hat{\sigma}_l^z |\uparrow\rangle = |\uparrow\rangle, \quad \hat{\sigma}_l^z |\downarrow\rangle = -|\downarrow\rangle, \tag{2.4}$$

and obey the commutation relation

$$\left[\hat{\sigma}_l^\alpha, \hat{\sigma}_l^\beta \right] = 2i\epsilon_{\alpha\beta\gamma} \hat{\sigma}_l^\gamma, \quad \alpha, \beta, \gamma \in \{x, y, z\} \tag{2.5}$$

among operators of the same site l, where for two operators $\hat{A}$ and $\hat{B}$ the commutator is defined as

$$[\hat{A}, \hat{B}] = \hat{A}\hat{B} - \hat{B}\hat{A}, \tag{2.6}$$

and the totally antisymmetric tensor $\epsilon_{\alpha\beta\gamma}$ is invariant under cyclic permutations of
its three indices, and assumes $\epsilon_{xyz} = 1$ and $\epsilon_{yxz} = -1$ while it is zero if two or more
indices are equal. Further, the Pauli operators commute for different sites $k \neq l$:

$$\left[\hat{\sigma}_l^\alpha, \hat{\sigma}_k^\beta \right] = 0, \quad \alpha, \beta \in \{x, y, z\}. $$

In this work we focus on quenches which involve different magnetic phases in the
state preparation and time evolution stage. For $\Delta > 1$ the ground state of (2.1) has
antiferromagnetic order and the model is said to describe a Néel-Ising-phase,
where a finite energy gap separates the ground from higher excited states.

Deep in the Néel Ising phase of the XXZ chain ($\Delta \gg 1$) an estimation for the
gap is obtained by realising that elementary excitations will be the states which
arise upon flipping one spin in the perfectly ordered antiferromagnet (Fig. 2.1).
This will affect the bonds left and right to that flipped spin in such a way that now

Fig. 2.1 Domain wall excitations in the Néel Ising phase of the XXZ model, which are separated by an energy gap $\sim 2J\Delta$ from the ground state. Figure reprinted from [14]

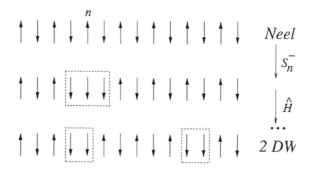

two *aligned* spins will reside on neighbouring lattice sites (so called domain walls). A suitable superposition of all states with two domain walls will be the first excited state, and the energy that is necessary to create two domain walls amounts to $\sim 2J\Delta$ in this limit [14]. The gap vanishes upon approaching $\Delta \rightarrow 1^{+}$, the isotropic point, where the Hamiltonian is fully rotational symmetric in spin space.

The parameter range $-1 < \Delta < 1$ corresponds to the gapless XY phase of the model. At $\Delta = 0$ the model becomes particularly simple in that can be mapped to a system of spinless fermions [15] which hop between lattice sites and interact only via the exclusion principle (no more than one fermion is allowed at each and every lattice site). For nonzero anisotropy Δ, depending on the sign, repulsive or attractive forces among the fermions on neighbouring lattice sites give rise to scattering effects making the solution substantially harder [16]. We will exploit the exact mapping to fermions in the case $\Delta = 0$ in (Sect. 2.4).

Finally, for anisotropies $\Delta < -1$ the ground state is two-fold degenerate and displays ferromagnetic order where spins tend to align parallel.

Since irrespective of the value of Δ one has that

$$\left[\hat{H}, \hat{S}_z\right] = 0, \tag{2.7}$$

with $\hat{S}_z = \sum_{l=1}^{N} \hat{\sigma}_l^z$, the ground state of $\hat{H}$ has is also an eigenvector of $\hat{S}_z$ with a definite value of the total z-magnetisation $\langle \hat{S}_z \rangle$. Furthermore, it is implied that $\langle \hat{S}_z \rangle$ is a constant of motion when the time evolution is generated by H.

2.2 Quantum Quench and Distant Entanglement Generation

Figure 2.2 pictures our scenario, where Alice and Bob are situated at opposite ends of a one dimensional (1D) lattice of perpetually interacting spin-$\frac{1}{2}$ particles. In this chapter we suggest a scheme which allows the establishment of a strong entanglement between Alice's and Bob's spins (the remotest spins of the lattice) without any requirement of local control for the preparation of the initial state of the chain or for the subsequent dynamics. In our scheme, first the lattice of strongly interacting spins is cooled to its Néel Ising ground state. Then, upon instantly changing the anisotropy

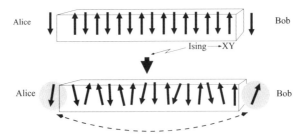

Fig. 2.2 Schematic of our proposal of entangling distant spins. Alice and Bob are at opposite ends of the chain. Upon cooling the system to the Ising ground state and subsequent non-adiabatic switching to an XY interaction entanglement is being established between Alice's and Bob's spins. Figure reprinted from [17]

parameter in its Hamiltonian, thereby crossing the phase border separating Néel Ising phase from the XY phase, the pair of edge spins evolve to a highly entangled mixed state. For this state, entanglement purification methods are known [18–20], which Alice and Bob can use to convert, only by local operations and classical communication, a few (say n) copies of the state to $m < n$ pure $|\psi^-\rangle$ states (see Sect. 2.5). These $|\psi^-\rangle$ could then be used to teleport any state from Alice to Bob.

In what follows, we will formulate the case of time-evolution of the Ising ground state under action of the XX Hamiltonian. This corresponds to an instantaneous, i.e. idealised quench in the anisotropy parameter $\Delta_1 \to \Delta_2$ with $\Delta_1 \to \infty$ and $\Delta_2 = 0$ thereby crossing critical value $\Delta = 1$, which separates the Néel-Ising-phase from the XY-phase. For $\Delta \gg J$ the Ising ground state gets arbitrarily close to the ideal Néel state, which is twofold degenerate in the absence of an external field. These ideal Néel states arise from the perfectly polarised state $|\Downarrow_N\rangle$ upon flipping every other spin:

$$|\mathcal{N}_1\rangle \equiv |\downarrow_1, \uparrow_2, \downarrow_3, \dots\rangle$$

and

$$|\mathcal{N}_2\rangle \equiv |\uparrow_1, \downarrow_2, \uparrow_3, \dots\rangle.$$

Note that these two states turn into each other by a spin flip at each place, i.e. $|\mathcal{N}_1\rangle = (\prod_{k=1}^{N} \hat{\sigma}_k^x)|\mathcal{N}_2\rangle$ and vice versa. In an experiment, the initial preparation of the Néel-Ising-ground state will yield, at low enough temperatures, an equal mixture of both Néel orders and negligible admixture of higher energy eigenstates. We adopt the notion of *thermal ground state* from [21] for

$$\hat{\rho}_0 = \frac{1}{2}(|\mathcal{N}_1\rangle\langle\mathcal{N}_1| + |\mathcal{N}_2\rangle\langle\mathcal{N}_2|), \tag{2.8}$$

which exhibits the same symmetries as the Ising Hamiltonian $H(\Delta \to \infty)$, as opposed to each individual, degenerate ground state of the antiferromagnetic Ising-chain.

We would like to investigate the dynamics of the long range nonclassical correlations, i.e. the entanglement between the first and the last spin of the chain, provided the system is initially prepared in the global state (2.8) and assuming $\Delta = 0$ for the subsequent time evolution. To this end, we compute the state $\hat{\rho}_{1,N}$ (reduced density operator) of the remotest pair of spins, from which we can deduce the amount of entanglement between them. We find that after an optimal time $T_{\max}$ the pair of edge spins evolves into a entangled mixed state, which for shorter chains is very well described by

$$\hat{\rho}_{1,N} \simeq f|\psi^+\rangle\langle\psi^+| + \frac{(1-f)}{2}(|\uparrow,\uparrow\rangle\langle\uparrow,\uparrow| + |\downarrow,\downarrow\rangle\langle\downarrow,\downarrow|),$$

where f is the *fully entangled fraction* which is a figure of merit in quantum communication theory and indicates whether a mixed entangled state is purifiable (see Sect. 2.5)

In the following sections, we will discuss the results that were summarised above in greater detail alongside a presentation of the methods that are involved in their derivation. To this end, in Sect. 2.3 we introduce the Hilbert–Schmidt operator decomposition which leads to the expression of $\hat{\rho}_{1N}$ in terms of correlation functions. Section 2.4 deals with the mapping of the spin degrees of freedom to those of fermions which allows us to solve the time dependence of the correlation functions. We also discuss the subject of entanglement distillation and the effects of disorder.

2.3 Calculation of Reduced Density Operators From Correlation Functions

The nonclassical part of correlations that is established between the terminal sites of the spin chain, labelled 1 and N, is encoded in the reduced density operator

$$\hat{\rho}_{1,N} \equiv \text{Tr}_{2\cdots N-1}[\hat{\rho}], \tag{2.9}$$

where $\hat{\rho}$ is the state of the total system which evolves unitarily under the action of the XXZ Hamiltonian (2.1). Instead of evaluating (2.9) explicitly, it is more convenient to express $\hat{\rho}_{1,N}$ in terms of two point correlation functions. In general, any operator $\hat{A}$ on the state space of the chain composed of N spin-$1/2$ particles may be decomposed into

$$\hat{A} = \sum_{k=1}^{2^{2N}} \text{Tr}\left[\hat{\rho}\,\hat{X}_k^\dagger\right]\hat{X}_k \tag{2.10}$$

$$\left(\hat{X}_k, \hat{X}_l\right) \equiv \text{Tr}\left[\hat{X}_k^\dagger\hat{X}_l\right] = \delta_{k,l} \tag{2.11}$$

$$\sum_k \hat{X}_k = \mathbb{1} \qquad (2.12)$$

which is sometimes referred to as Hilbert–Schmidt operator decomposition [22]. For an operator A defined on the Hilbert space spanned by the two spin-$\frac{1}{2}$ degrees of freedom associated to site l of a spin chain configuration it is convenient to choose the following set of orthogonal operators:

$$\{\hat{X}\} = \left\{\hat{\sigma}_l^+, \hat{\sigma}_l^-, \hat{P}_l^\uparrow, \hat{P}_l^\downarrow\right\}, \quad l = 1, 2, \ldots, N \qquad (2.13)$$

$$\hat{\sigma}_l^\pm = \frac{1}{2}(\hat{\sigma}_l^x \pm i\hat{\sigma}_l^y) \qquad (2.14)$$

$$\hat{P}_l^\downarrow = \hat{\sigma}_l^- \hat{\sigma}_l^+ \qquad (2.15)$$

$$\hat{P}_l^\uparrow = \hat{\sigma}_l^+ \hat{\sigma}_l^-. \qquad (2.16)$$

An operator basis for the entire spin chain is obtained by means of a direct product of these operators on different sites in all possible combinations. With respect to the computational basis $\{|\uparrow\uparrow\rangle, |\uparrow\downarrow\rangle, |\downarrow\uparrow\rangle, |\downarrow\downarrow\rangle\}$ the following representation for $\hat{\rho}_{1,N}$ is obtained:

$$\hat{\rho}_{1,N} = \begin{pmatrix} \langle \hat{P}_1^\uparrow \hat{P}_N^\uparrow \rangle & \langle \hat{P}_1^\uparrow \hat{\sigma}_N^- \rangle & \langle \hat{\sigma}_1^- \hat{P}_N^\uparrow \rangle & \langle \hat{\sigma}_1^- \hat{\sigma}_N^- \rangle \\ \langle \hat{P}_1^\uparrow \hat{\sigma}_N^+ \rangle & \langle \hat{P}_1^\uparrow \hat{P}_N^\downarrow \rangle & \langle \hat{\sigma}_1^- \hat{\sigma}_N^+ \rangle & \langle \hat{\sigma}_1^- \hat{P}_N^\downarrow \rangle \\ \langle \hat{\sigma}_1^+ \hat{P}_N^\uparrow \rangle & \langle \hat{\sigma}_1^+ \hat{\sigma}_N^- \rangle & \langle \hat{P}_1^\downarrow \hat{P}_N^\uparrow \rangle & \langle \hat{P}_1^\downarrow \hat{\sigma}_N^- \rangle \\ \langle \hat{\sigma}_1^+ \hat{\sigma}_N^+ \rangle & \langle \hat{\sigma}_1^+ \hat{P}_N^\downarrow \rangle & \langle \hat{P}_1^\downarrow \hat{\sigma}_N^+ \rangle & \langle \hat{P}_1^\downarrow \hat{P}_N^\downarrow \rangle \end{pmatrix} \qquad (2.17)$$

One can exploit the intrinsic symmetries of H to reduce the number of non-zero matrix entries above considerably. For states with fixed z-magnetisation (including the ground state of H, as can be inferred from (2.7)) the following expectation values along with their complex conjugates must vanish:

$$\langle \hat{P}_1^\uparrow \hat{\sigma}_N^- \rangle = \langle \hat{\sigma}_1^- \hat{P}_N^\uparrow \rangle = \langle \hat{\sigma}_1^- \hat{\sigma}_N^- \rangle = \langle \hat{\sigma}_1^- \hat{P}_N^\downarrow \rangle = \langle \hat{P}_1^\downarrow \hat{\sigma}_N^- \rangle = 0.$$

Further, the remaining off-diagonal elements $\langle \hat{\sigma}_1^- \hat{\sigma}_N^+ \rangle$ and $\langle \hat{\sigma}_1^+ \hat{\sigma}_N^- \rangle$ must be real owing to the reflection symmetry about the middle of the chain. Hence, $\hat{\rho}_{1,N}$ has only the following non-zero elements:

$$\hat{\rho}_{1,N} = \begin{pmatrix} a & & & \\ & b & c & \\ & c & b & \\ & & & a \end{pmatrix} \qquad (2.18)$$

The following section deals with the question of how to compute the time dependent quantities a, b, c, d for our specific scenario.

2.4 Mapping to a Lattice Fermion Model

We will now seek for a solution to the equations of motion governing the time dependence of the reduced density operator (2.9) after an instantaneous quench from an initial $\Delta = \infty$ to a final $\Delta = 0$ was performed at time $t = 0$. For the model under consideration, this is customarily done by passing over from spin degrees of freedom to fermionic ones. By introducing Jordan-Wigner fermions

$$\hat{c}_l^{\dagger} \equiv \left(\prod_{n=1}^{l-1} -\hat{\sigma}_l^z \right) \hat{\sigma}_l^{+} \tag{2.19}$$

which obey the canonical anticommutation relations (CAR)

$$[\hat{c}_k^{\dagger}, \hat{c}_l]_{+} = \delta_{k,l}, \tag{2.20}$$

$$[\hat{c}_k^{\dagger}, \hat{c}_l^{\dagger}]_{+} = [\hat{c}_k, \hat{c}_l]_{+} = 0, \tag{2.21}$$

where the anticommutator is defined as $[\hat{A}, \hat{B}]_{+} = \hat{A}\hat{B} + \hat{B}\hat{A}$, the XXZ Hamiltonian at $\Delta = 0$ (XX model, henceforth) turns into to a model of free[1] fermions

$$\hat{H}(\Delta = 0) = J \sum_{l=1}^{N-1} \hat{c}_l^{\dagger} \hat{c}_{l+1} + \hat{c}_{l+1}^{\dagger} \hat{c}_l. \tag{2.22}$$

The matrix elements of $\hat{\rho}_{1N}$ can be worked out in this language by diagonalising (2.22). We show in Appendix C that in the Heisenberg picture

$$a = \langle \hat{c}_1^{\dagger}(t)\hat{c}_1(t) \rangle_1 \langle \hat{c}_N^{\dagger}(t)\hat{c}_N(t) \rangle_1 - \langle \hat{c}_1^{\dagger}(t)\hat{c}_N(t) \rangle_1 \langle \hat{c}_N^{\dagger}(t)\hat{c}_1(t) \rangle_1$$
$$- \frac{1}{2} \left(\langle \hat{c}_1^{\dagger}(t)\hat{c}_1(t) \rangle_1 + \langle \hat{c}_N^{\dagger}(t)\hat{c}_N(t) \rangle_1 - 1 \right), \tag{2.23}$$

$$b = \frac{1}{2} - a, \tag{2.24}$$

$$c = \frac{1}{2} \left((-1)^{M+1} \langle \hat{c}_N^{\dagger}(t)\hat{c}_1(t) \rangle_1 + c.c. \right). \tag{2.25}$$

Above we introduced the shorthand notation $\langle \cdots \rangle_1 = \langle \mathcal{N}_1 | \cdots | \mathcal{N}_1 \rangle$ and M is the conserved number of spin up states in the dynamical Néel state

$$\left(\sum_{l=1}^{N} \hat{P}_l^{\uparrow} \right) e^{-i\hat{H}t} |\mathcal{N}_1\rangle = M e^{-i\hat{H}t} |\mathcal{N}_1\rangle ,$$

---

[1] The term free refers to the fact that the Hamiltonian may be cast into diagonal form of noninteracting fermionic degrees of freedom, see Appendix A.

i.e. $M = N/2$ for even N and $M = (N - 1)/2$ for odd N. The Heisenberg picture operators $\hat{c}_k^\dagger(t)$ read

$$\hat{c}_k(t) = e^{iHt}\hat{c}_k e^{-iHt} = \sum_{l=1}^{N} f_{k,l}(t)\hat{c}_l \tag{2.26}$$

with

$$f_{k,l}(t) \equiv \sum_{m=1}^{N} g_{k,m} g_{m,l} e^{i\epsilon_m t} \tag{2.27}$$

$$\begin{aligned} g_{k,l} &= \sqrt{\frac{2}{N+1}} \sin(q_k l), \\ \epsilon_k &= 2J \cos(q_k), \\ q_k &= \frac{\pi k}{N+1}, \end{aligned} \tag{2.28}$$

whereby the two point correlation function occurring in (2.23)–(2.25) assume

$$\langle \hat{c}_i^\dagger(t)\hat{c}_j(t)\rangle_1 = \sum_{k=1}^{N} f_{i,k}(t) f_{j,k}^*(t) \ \langle \hat{c}_k^\dagger(0)\hat{c}_k(0)\rangle_1 = \sum_l f_{i,2l}(t) f_{j,2l}^*(t) \tag{2.29}$$

where we used that at $t = 0$

$$\langle \mathcal{N}_1 | \hat{c}_k^\dagger(0)\hat{c}_k(0) | \mathcal{N}_1 \rangle = \delta_{k,2l}, \ l = 1, 2, \ldots. \tag{2.30}$$

Figure 2.3 shows the time evolution of the matrix elements for a chain of $N = 9$ spins.

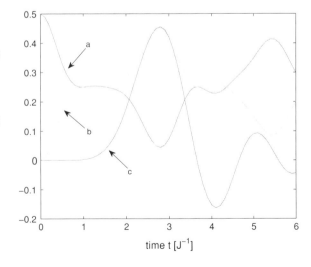

Fig. 2.3 Time evolution after quench for $N = 9$. By means of numerical search one finds that after an optimal time $T_{max} = 2.7905J^{-1}$ the matrix elements of $\hat{\rho}_{1,N}$ assume $a = 0.0433$, $b = 0.4567$, $c = 0.4550$ which amounts to a purifiable mixed state

time t [J^{-1}]

2.5 Entanglement Purification

We showed in the preceding sections that the state of the pair of edge spins $\hat{\rho}_{1,N}$ in the open XXZ chain is a mixed state and assumes the form (2.18) during time evolution after a quench which is performed within this model class (i.e., Δ is the parameter which triggers the quench). Let us now discuss the circumstances under which $\hat{\rho}_{1,N}$ can be considered a useful resource for quantum communication tasks. While it has become customary to quantify mixed state entanglement of two qubits by means of concurrence [23], we will adopt for the following discussion a more transparent measure, the fully entangled fraction f, which has an immediate operational interpretation in terms of *entanglement purification*.

That is, we ask whether $\hat{\rho}_{1,N}$ contains sufficient entanglement between the qubits for it to be *purifiable* into maximally entangled pure states and, if so, how efficiently this can be done.

The density matrix (2.18) can be understood as a mixture of Bell states of pairs of qubits

$$|\psi^{\pm}\rangle = \frac{1}{\sqrt{2}}(|\uparrow\downarrow\rangle \pm |\downarrow\uparrow\rangle) \tag{2.31}$$

$$|\phi^{\pm}\rangle = \frac{1}{\sqrt{2}}(|\uparrow\uparrow\rangle \pm |\downarrow\downarrow\rangle). \tag{2.32}$$

The Bell states span the two-qubit Hilbert space, and in this representation $\hat{\rho}_{1,N}$ is diagonal:

$$\hat{\rho}_{1,N} = (b-c)\,|\psi^{-}\rangle\langle\psi^{-}| + (b+c)\,|\psi^{+}\rangle\langle\psi^{+}| + a\left(|\phi^{+}\rangle\langle\phi^{+}| + |\phi^{-}\rangle\langle\phi^{-}|\right). \tag{2.33}$$

Assume that two parties Alice (A) and Bob (B) are supplied with n copies of the state $\hat{\rho}_{1,N}$ of a pair of qubits and that the parties have access to one member of these pairs respectively. The object of entanglement purification is to extract a lesser number $m < n$ of asymptotically pure, maximally entangled states from this initial supply only by local operations and classical communication (LOCC). A procedure for this purpose, the so called recurrence method, has been proposed in [18, 19]. Since its inception many refined methods have been developed [24], yet the core idea is the aggregation of the weight of one of the Bell states in the above mixture at the expense of the weights of the others through LOCC. The fist step of the recurrence method is to draw two impure pairs of qubits in state $\hat{\rho}_{1,N}$ from the initial supply. Subsequently, these pairs are entangled by applying local unitary operations leading to a permutation of basis states $|\psi^{\pm}\rangle, |\phi^{\pm}\rangle$ and to alterations in their corresponding statistical weights. A local measurement in the $\{|\uparrow\rangle, |\downarrow\rangle\}$ basis on one of the pairs leaves the other, untouched pair in a state which can be, statistically speaking, less mixed and more entangled. While the measured pair becomes expendable, Alice and Bob can infer from comparing their measurement results whether the untouched pair is to be kept or to be discarded. The subset of kept pairs serves as the set of copies for the subsequent iteration of the recurrence method, and will—under suitable conditions—gradually be approaching a set of almost pure, maximally entangled states of pairs of qubits.

In the general case of an arbitrary density operator, a figure of merit for how well purification methods can perform is the *fully entangled fraction f* defined as

$$f(\hat{\rho}) = \max_{|e\rangle} \langle e|\hat{\rho}|e\rangle \tag{2.34}$$

where $\hat{\rho}$ is an arbitrary mixed state of two qubits and the maximisation is carried out over the set of all maximally entangled states $|e\rangle$. In a state $\hat{\rho}$ which is diagonal in the basis of Bell states, f is given by its maximum eigenvalue [20], hence for $\hat{\rho}_{1,N}$ of (2.33) one has

$$f(\hat{\rho}_{1,N}) = \max(b + |c|, a). \tag{2.35}$$

A density operator $\hat{\rho}$ of two qubits is called *purifiable* if [18, 19]

$$f(\hat{\rho}) > \frac{1}{2}$$

which is sufficient for the recurrence method to have a finite yield. The yield is the fraction $\frac{m}{n}$ of pairs in state $\hat{\rho}$ with individual

$$f(\hat{\rho}) = 1 - \epsilon, \quad 0 \le \epsilon < \frac{1}{2}$$

which can be obtained, on average, from $n > m$ initial pairs after a finite number of iterations of the recurrence method. The yield goes to zero as ϵ goes to zero [24], i.e. the recurrence method can achieve pure and maximally entangled state in the asymptotic sense. In our specific case, $\hat{\rho}_{1,N}$ is purifiable if

$$b + |c| > \frac{1}{2}. \tag{2.36}$$

2.6 Results and Discussion

The idealised quench discussed so far gives rise to a significant amount of entanglement at an optimal time $T_{\max}$ after the instant of quench, in that in this model it is purifiable ($f > 0.5$) for chains of up to $N = 241$ spins (Fig. 2.4). We observe a linear scaling of $T_{\max}$ with system size, and the peak value of fully entangled fraction f vanishes according to a power-law $\sim N^{-0.22(9)}$ for $N > 20$.

In addition to this quench ($\Delta : \infty \to 0$), we have studied more general quenches purely numerically for small system sizes. We studied the effects of (a)

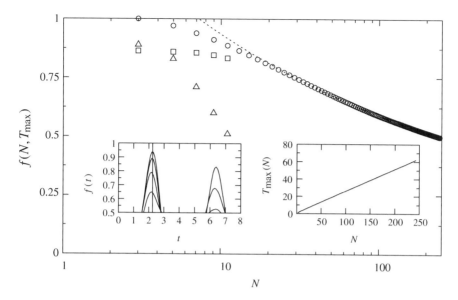

Fig. 2.4 Fully entangled fraction f as a function of the number of lattice sites N in a semi-logarithmic scale. The data points for the analytic quench scenario (circles) exceed 0.5 for chains up to $N = 241$ spins. We find a good agreement of these data for $N \geq 25$ to a function of the form $g^{\text{fit}}(N) \propto N^{-\nu}$ with $\nu = 0.22(9)$ (dashed curve). The main panel is supplemented with data from numerical study of more general quenches and small system sizes $N \leq 11$. Triangles refer to a quench $\Delta : \infty \to 1$ and squares label data for a quench $\Delta : 3 \to 0$ *Left inset*. Typical time evolution of the fully entangled fraction after quench $\Delta : \infty \to 0$ for different values of random disorder parameter ($\delta = 0, 0.1, 0.2, 0.3$ for top to *bottom curves*, respectively). The lines refer to an average taken over 100 independent realisations of disorder. *Right inset:* Linear size-scaling of the optimal time $T_{\max}$ (circles), which is needed for f to reach its first maximum value. Figure reprinted from [17]

having an initial thermal ground state corresponding to finite Δ which deviates from the ideal mixture of the two Néel states (see (2.8)) and (b) time evolution being generated by a Hamiltonian with $0 < \Delta \leq 1$. Numerics further allows us to introduce random couplings which constitutes a form of disorder, subject to which entanglement becomes degraded to a certain extent. Random couplings amounts to choosing a site dependent coupling strength in the XXZ Hamiltonian (2.1)

$$J \to J_l = J(1 + \delta_l)$$

with normally distributed random numbers δ_l having zero mean and standard deviation δ. Figure 2.4 summarises our results.

One recognises from the data shown in the main panel of Fig. 2.4 that it is not as crucial to have an ideal Néel-type ground state before the quench as it is to time evolve with the XX-Hamiltonian (which corresponds to the free fermion case). Therefore the dispersion-less and scattering-free evolution, that is expected for the excitations of the XX spin chain, is seen to be the most important ingredient, at least for longer chains, for purifiable entanglement to be established. Still, over short ranges more general quenches obey the qualitative features that are heralded by the idealised quench, and could be of interest for experiments where only a few coupled qubits are involved.

2.6.1 Qualitative Origin of Entanglement Growth

The linear scaling of $T_{\max}$ with system size N, as shown in the lower right panel, suggests that entanglement displays *causality*. Distant correlations are therefore very likely due to earlier local events which tend to spread information with a constant speed across the lattice. Calabrese and Cardy first proposed [3] this idea of causality in the context of entanglement in many-body systems. They argued that, due to quench, counter propagating quasiparticle excitations are emitted from each lattice site in a spin chain and contribute a certain amount of entanglement between a region of the chain of length l and its complement once their causal cones (a pictorial notion of left and right moving wave-fronts as functions of time) admit this, i.e. once one of the particles (but not both) is contained within the region. With this in mind, one might expect that in our quench scenario entanglement is established roughly at the time when a pair of right and left moving particles, emitted from the middle of the chain, have simultaneously reached both ends. Given that the Fermi velocity of excitations in the XX model reads as

$$v_F = \left| \left(\partial_{q_k} \epsilon_k \right) \big|_{q_k = \frac{\pi}{2}} \right| = 2J,$$

one would expect that the time at which entanglement is first optimally established will roughly scale as $T_{\max} \sim \frac{N}{2v_F} = \frac{N}{4J}$. However, our data for $T_{\max}$ in a quench $\Delta : \infty \to 0$ are seen to obey the scaling $T_{\max} \sim \frac{N}{\pi J}$ suggesting that the process of establishing entanglement is slightly slower than this estimate. Yet, the linear scaling with N is otherwise consistent with the causal cone picture.

This delay hints towards a collective effect, leading to an optimal entanglement as a result of many such local events. The excitation that is emitted from the centre of the chain (meaningful in the case of odd N) will, at the optimal time $T_{\max}$, have encountered the terminal sites and will have been reflected from the boundaries. Excitations which are emitted far from the middle, say, to the left of the middle are yet to encounter the right end, and have been reflected from the left boundary (Fig. 2.5).

A subtlety arising in this study is that quenches in chains of even numbers of spins do not give rise to purifiable entanglement at any time at all. Only if one starts form pure initial states $|\mathcal{N}_1\rangle$ or $|\mathcal{N}_2\rangle$ even and odd chains behave equivalently and time evolution produces entanglement between the ends in either case. Hence, the mixedness must be made responsible for this even/odd effect. We do not have a conclusive explanation for this phenomenon at present.

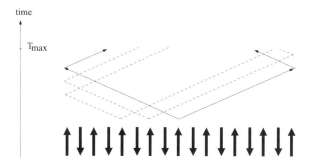

Fig. 2.5 The establishment of entanglement as a collective effect. Quasiparticle (*quasi hole*) excitations are emitted locally from each $|\uparrow\rangle$ ($|\downarrow\rangle$), and give rise to a maximal fidelity with $|\psi^+\rangle$ at time T_{max}, which occurs slightly later than the instance of time where wave-fronts which were emitted from the middle hit the edges (*solid arrows*). We expect, that this is due to the finite width of each wave-packet, which travels rather dispersion-less (in the idealised quench) through the lattice. The edge spins then encounter contributions from multiple excitations, sources of which are located in a broader region in the vicinity of the middle of the chain (indicated by *dashed lines*)

2.6.2 Thermalisation and Decoherence

As to the preparation of the initial antiferromagnetically ordered Néel state, we assumed a statistical mixture of the degenerate ground states thereby taking into account thermalisation to a certain extent. Neglecting admixture of first and higher excited states will be appropriate if the energy gap $|E_1 - E_0|$ between ground state and first excited is large compared to $k_B T$ where k_B is the Boltzmann constant and T is the temperature. We argued in Sect. 2.1 that the gap of the Ising antiferromagnet is proportional to the coupling strength J. Therefore strong coupling is preferable in order to prevent the system from thermalising.

The present study neglects effects which may stem from an interaction of the spin chain to a thermal environment *during* the time evolution after quench. This could be a valid assumption given that the process is essentially rather fast: The optimal time after which entanglement is first maximally established scales linearly in the system size N (see Fig. 2.4) and is inverse proportional to the coupling strength. It is therefore hoped that under suitable conditions the unitary evolution can be sufficiently fast for decoherent effects to be negligible.

2.6.3 Possible Experimental Implementations

The prospect of evidencing entanglement in a strongly correlated quantum system using our quench based entangling scheme seems promising regarding recent experiments in different setups. Suitable candidates for experimental tests will be briefly reviewed hereafter.

Cold atoms in artificial lattice potentials are an ideal testbed to observe of coherent quantum evolution after a quench [9]. In particular, the preparation of a chain of atoms in a magnetic Néel state which our quench protocol relies on (more precisely the pure state $|\mathcal{N}_1\rangle$ or $|\mathcal{N}_2\rangle$) has been realised with high fidelity in a recent experiment [10], which also provided evidence for effective spin interactions between neighbouring atoms. Yet, it might be challenging to evidence entanglement between the atoms of particular sites including those at the ends. This would require full state tomography of the ending spins.

Ions trapped by laser light [25] constitute another example of a very *clean* system of strongly interacting particles, which can be cooled, confined and coherently manipulated reliably for long time while experiencing only little perturbation from the environment [26]. The relevant degrees of freedom for a collection of trapped ions are the individual internal spin states, as well as their collective motional modes which are caused by Coulomb forces. Spin and vibrational degrees of freedom can be coupled via incident laser light giving rise to an effective spin-spin interaction among the ions [27]. While entangled states of pairs of ions can be reliably obtained by other means, it will still be interesting to check the predictions of our scheme in this highly controllable environment.

A simple architecture involving superconductor circuits is one where Josephson junctions—formed by two superconductors separated by a thin, insulating layer— form a linear array. Two quantum conjugate variables which can be used to form a suitable qubit are the number of electron (Cooper) pairs on the electrode of a junction giving rise to charge state qubits [28] or the phase difference between the wave functions which arises upon crossing the insulating barrier and can be used to form a flux qubit [29]. Coherent quantum evolution in superconducting circuits were observed [30] and the dephasing time was reported to be on the order of 20 ns, while in recent experiments the μs regime is approached. Coupling different units to form chains or more elaborate architectures as well as switching the interactions dynamically is subject of ongoing experimental efforts [31]. Methods of simulating effective spin Hamiltonians in chain like architectures of Josephson junctions were proposed in [32, 33]. A notable advantage is the mesoscopic scale of these solid state devices which would enable single site access for measurement and control. A noteworthy drawback is that highly elaborate cryogenic techniques are required to cool the circuits to the mK regime [30]. Together with fabrication this makes these experiments quite costly.

2.7 Related Work

Galve et al. [34] proposed a similar scheme to entangle the distant ends of a spin chain by periodically driving the coupling strengths $J = J(t)$ in a XY spin chain with a transverse magnetic field. There, the initially uncoupled spin system, i.e. $J(0)$, is cooled to the ferromagnetic ground state by virtue of a strong external magnetic field. A periodic driving $J(t) \sim \sin(\omega_d t)$ subsequently leads to a dynamical response of the spin chain and, moreover, gives rise to a significant amount of entanglement between the edge spins, particularly at resonance, i.e. where ω_d matches the Zeeman energy splitting of each spin. In this resonant case the dynamical driving is seen to be mathematically equivalent to the static quench described in this chapter [34].

References

1. K. Sengupta, S. Powell, S. Sachdev, Quench dynamics across quantum critical points. Phys. Rev. A **69**(5), 053616 (2004)
2. M. Cramer, C.M. Dawson, J. Eisert, T.J. Osborne, Exact relaxation in a class of nonequilibrium quantum lattice systems. Phys. Rev. Lett. **100**(3), 030602 (2008)
3. P. Calabrese, J. Cardy, Evolution of entanglement entropy in one-dimensional systems. J. Stat. Mech. Theory Exp. **2005**(04), P04010 (2005)
4. G. De Chiara, S. Montangero, P. Calabrese, R. Fazio, Entanglement entropy dynamics of heisenberg chains. J. Stat. Mech. Theory Exp. **2006**(03), P03001 (2006)
5. L. Cincio, J. Dziarmaga, M.M. Rams, W.H. Zurek, Entropy of entanglement and correlations induced by a quench: dynamics of a quantum phase transition in the quantum ising model. Phys. Rev. A **75**(5), 052321 (2007)
6. M. Greiner, O. Mandel, T. Esslinger, T.W. Hansch, I. Bloch, Quantum phase transition from a superfluid to a mott insulator in a gas of ultracold atoms. Nature **419**, 51 (2002)
7. T. Kibble, Phase-transition dynamics in the lab and the universe. Phys. Today **60**(9), 47–52 (2007)
8. O. Kinoshita, T. Wenger, D.S. Weiss, A quantum newton's cradle. Nature **440**, 900 (2006)
9. I. Bloch, Quantum gases. Science **319**(5867), 1202–1203 (2008)
10. S. Trotzky, P. Cheinet, S. Folling, M. Feld, U. Schnorrberger, A.M. Rey, A. Polkovnikov, E.A. Demler, M.D. Lukin, I. Bloch, Time-resolved observation and control of superexchange interactions with ultracold atoms in optical lattices. Science **319**(5861), 295–299 (2008)
11. A.B. Kuklov, B.V. Svistunov, Counterflow superfluidity of two-species ultracold atoms in a commensurate optical lattice. Phys. Rev. Lett. **90**(10), 100401 (2003)
12. L.-M. Duan, E. Demler, M.D. Lukin, Controlling spin exchange interactions of ultracold atoms in optical lattices. Phys. Rev. Lett. **91**(9), 090402 (2003)
13. E. Altman, A. Auerbach, Oscillating superfluidity of bosons in optical lattices. Phys. Rev. Lett. **89**(25), 250404 (2002)
14. H.-J. Mikeska, A.K. Kolezhuk, *Lecture Notes in Physics: Quantum Magnetism* , vol. 646 (Springer, Berlin, 2004) pp. 1–83
15. E. Lieb, T. Schultz, D. Mattis, Two soluble models of an antiferromagnetic chain. Ann. Phys. **16**(3), 407–466 (1961)
16. T. Giamarchi, Quantum Physics in One Dimension. (Clarendon Press, Oxford) 2003
17. H. Wichterich, S. Bose, Exploiting quench dynamics in spin chains for distant entanglement and quantum communication. Phys. Rev. A **79**(6), 060302(R) (2009)
18. C.H. Bennett, G. Brassard, S. Popescu, B. Schumacher, J.A. Smolin, W.K. Wootters, Purification of noisy entanglement and faithful teleportation via noisy channels, Phys. Rev. Lett. **76**, 722 (1996)
19. C.H. Bennett, G. Brassard, S. Popescu, B. Schumacher, J.A. Smolin, W.K. Wootters, Purification of noisy entanglement and faithful teleportation via noisy channels, Phys. Rev. Lett. **78**(10), 2031 (1997)
20. C.H. Bennett, D.P. DiVincenzo, J.A. Smolin, W.K. Wootters, Mixed-state entanglement and quantum error correction. Phys. Rev. A **54**(5), 3824–3851 (1996)

21. T.J. Osborne, M.A. Nielsen, Entanglement in a simple quantum phase transition. Phys. Rev. A **66**(3), 032110 (2002)
22. H.-P. Breuer, F. Petruccione, The theory of open quantum systems. (Clarendon Press, Oxford, UK, 2002)
23. W.K. Wootters, Entanglement of formation of an arbitrary state of two qubits. Phys. Rev. Lett. **80**(10), 2245–2248 (1998)
24. D. Bruß, G. Leuchs, Lectures on quantum information. (Weinheim, Wiley-VCH, 2007)
25. J.I. Cirac, P. Zoller, Quantum computations with cold trapped ions. Phys. Rev. Lett. **74**(20), 4091–4094 (1995)
26. B. Wineland, Nature, **453**(19), 1008 (2008)
27. D. Porras, J.I. Cirac, Effective quantum spin systems with trapped ions. Phys. Rev. Lett. **92**(20), 207901 (2004)
28. Y. Nakamura, Y.A. Pashkin, J.S. Tsai, Coherent control of macroscopic quantum states in a single-cooper-pair box. Nature **398**, 786–788 (1999)
29. J.R. Friedman, V. Patel, W. Chen, S.K. Tolpygo, J.E. Lukens, Quantum superposition of distinct macroscopic states. Nature **406**, 43–46 (2000)
30. I. Chiorescu, Y. Nakamura, C.J.P.M. Harmans, J.E. Mooij, Coherent quantum dynamics of a superconducting flux qubit. Science **299**(5614), 1869–1871 (2003)
31. A.O. Niskanen, K. Harrabi, F. Yoshihara, Y. Nakamura, S. Lloyd, J.S. Tsai, Quantum coherent tunable coupling of superconducting qubits. Science **316**(5825), 723–726 (2007)
32. D. Giuliano, P. Sodano, Nucl. Phys. B **711**, 480 (2005)
33. A. Lyakhov, C. Bruder, Quantum state transfer in arrays of flux qubits. New J. Phys. **7**(1), 181 (2005)
34. F. Galve, D. Zueco, S. Kohler, E. Lutz, P. Haenggi, Phys. Rev. A **79**, 032332 (2009)

Chapter 3
Extraction of Pure Entangled States From Many-Body Systems by Distant Local Projections

In this chapter, we explore the feasibility of extracting a pure entangled state of noncomplementary and potentially well separated regions of a quantum many-body system. By studying the effect of measurements of local observables, such as the individual ground state magnetisation of separated blocks of spins of certain spin chains, we find that such an extraction can be accomplished, in principle, even though the state is a mixed state before the measurement. A general procedure is presented which is capable of predicting the optimal performance of extracting pure entangled state through local projective measurements. Numerical results for the ground state of the transverse XY spin ring suggest a connection of the projectively extractable pure entanglement $(\mathcal{E}_{PP})$ to the symmetries of the underlying model.

After a brief recap on the theory of ideal quantum measurements in Sect. 3.1, we introduce the notion of projectively extractable pure entanglement in Sect. 3.2 We illustrate the concept of $\mathcal{E}_{PP}$ on a simple example, namely that of a supersinglet state, in Sect. 3.3 We proceed by discussing a general procedure (Sect. 3.4) that predicts the optimal amount of pure entanglement which can be extracted on average from a general mixed state of two parties by way of local measurements. In the subsequent Sect. 3.5 the transverse XY spin chain will be introduced. This model will serve as a testbed for the general procedure, by looking at the pure entanglement that is extractable from the ground state. The results of this investigation are reported in Sect. 3.6

The work in this chapter has been done in collaboration with Dr. Javier Molina (Universidad de Cartagena, Spain), Prof. Vladimir Korepin (Stony Brook University, USA) and with my supervisor Prof. Sougato Bose.

3.1 Ideal Quantum Measurement

The von Neumann-Lüders measurement postulate may be formulated as follows [1, 2]. Quantum theory assigns to each measurable physical quantity X a Hermitian

H.C. Wichterich, *Entanglement Between Noncomplementary Parts of Many-Body Systems*, Springer Theses, DOI: 10.1007/978-3-642-19342-2_3, © Springer-Verlag Berlin Heidelberg 2011

Localisable entanglement

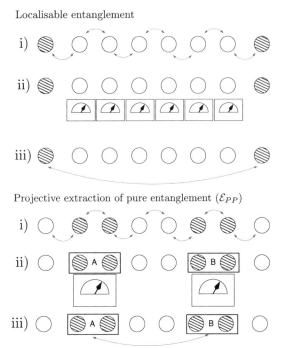

Fig. 3.1 Projective extraction of pure entanglement [3] versus localisable entanglement [4]. *Top panel*. Localisable entanglement (LE). In a typical quantum many-body system (i), entanglement is dominantly shared between neighbouring particles [5, 6]. By locally measuring observables of a designated subset of particles (ii) the range of entanglement among the other particles (depicted here as the pair of particles at the terminal sites of an open ended chain) can be increased. LE is the maximum amount of entanglement that can be localised between this pair, on average. *Bottom panel*. The initially unperturbed state of the many-body system (i) becomes subjected to measurements of observables that are local to the individual region of particles, designated A and B (ii). The post-measurement state of the particles in these regions can, under suitable conditions, be a pure and entangled state (iii). Projectively extractable pure entanglement $(\mathcal{E}_{PP})$ is the optimal amount of entanglement that can be extracted, on average, from the blocks by means of such projective measurements. Reproduced from [3]

operator $\hat{X}$, which is called observable (Fig. 3.1). Here, we consider observables with discrete spectral decomposition

$$\hat{X} = \sum_k x_k \hat{P}_k \tag{3.1}$$

where x_k is an eigenvalue of the observable $\hat{X}$ and corresponds to a possible measurement outcome. With each eigenvalue x_k one associates a projector

$$\hat{P}_k = \sum_{j_k=1}^{d_k} |x_k\rangle\langle x_k| \quad \Leftrightarrow \quad \hat{X}|x_k\rangle = x_k|x_k\rangle \tag{3.2}$$

where the sum accounts for a d_k-fold degeneracy of x_k. A projector is a Hermitian $\hat{P}_k^\dagger = \hat{P}$ and idempotent $\hat{P}^2 = \hat{P}$ operator.

Let $\hat{\rho}$ denote the density operator which describes a quantum statistical ensemble before a measurement has been carried out. Then, upon measuring the observable $\hat{X}$, which we suppose yields the outcome x_k, a new density operator

$$\hat{\rho}' = \frac{\hat{P}_k \hat{\rho} \hat{P}_k}{\mathrm{Tr}\left[\hat{P}_k \hat{\rho} \hat{P}_k\right]} \tag{3.3}$$

describes a sub-ensemble which is commensurate with this measurement outcome. The denominator in (3.3) ensures normalisation and coincides with the probability of obtaining the measurement outcome x_k

$$p(x_k) = \mathrm{Tr}\left[\hat{P}_k \hat{\rho} \hat{P}_k\right] = \mathrm{Tr}\left[\hat{P}_k \hat{\rho}\right] \tag{3.4}$$

where we used the cyclic property of the trace, $\mathrm{Tr}\left[\hat{A}\hat{B}\right] = \mathrm{Tr}\left[\hat{B}\hat{A}\right]$, and the i-dempotence of the projector. With a single projector $\hat{P}_k$ we can associate a *selective measurement* of observable $\hat{X}$, where all outcomes but x_k are discarded.

If the eigenvalue x_k is non-degenerate, then the mentioned sub-ensemble is describable by a pure state $|\Phi'\rangle$, even if $\hat{\rho}$ is a true mixture [2]. The circumstances by which the sub-ensemble is describable by a pure state $|\Phi'\rangle$, even though x_k is degenerate, will be laid out in the following section.

3.2 Extraction of Pure Entangled States by Distant Local Projective Measurements

Assume that a quantum system is comprised of two regions A and B and their complement $\overline{AB}$, where the regions could be spatially separated. We will denote the union of the two regions by $AB = A \cup B$.

Building on the theory of measurement laid out in the foregoing chapter, we ask whether it is feasible to perform a *selective local measurement* on the parts A and B of the form $\hat{P} = \hat{P}^A \otimes \hat{P}^B$ with outcome x that leads to a pure state $|\Phi\rangle$ after measurement which is also entangled with respect to the bipartition $A|B$. As a measure which captures both the amount of pure entanglement and the probabilistic nature of actually obtaining x we introduce the projectively extractable pure entanglement [3]

$$\mathcal{E}_{PP} = \max_{\left\{\hat{P}=\hat{P}_A \otimes \hat{P}_B | \hat{P}\hat{\rho}\hat{P}=p^{-1}|\Phi\rangle\langle\Phi|\right\}} \left\{p\,\mathcal{E}(|\Phi\rangle)\right\} \tag{3.5}$$

where $p = \mathrm{Tr}\left[\hat{P}\,\hat{\rho}\right]$ is the probability of measuring x.

Let us elaborate on the criteria that must be met so that a measurement gives rise to a pure state of the measured regions. Suppose that before the measurement the state of the whole system is described by a pure state

$$|\psi\rangle \in \mathcal{H}_A \otimes \mathcal{H}_B \otimes \mathcal{H}_{\overline{AB}}.$$

The Schmidt decomposition (Sect. 1.1.2) with respect to the bipartition $AB|\overline{AB}$ reads as

$$|\psi\rangle = \sum_{\alpha=1}^{\chi} \sqrt{w_\alpha} |w_\alpha^{AB}\rangle \otimes |w_\alpha^{\overline{AB}}\rangle \tag{3.6}$$

and the state of the subsystem, composed of the union of the regions AB, becomes

$$\hat{\rho} = \sum_{\alpha=1}^{\chi} w_\alpha |w_\alpha^{AB}\rangle\langle w_\alpha^{AB}| \tag{3.7}$$

By virtue of (3.3) one has

$$\hat{\rho}' = p^{-1} \sum_{\alpha} w_\alpha \hat{P} |w_\alpha^{AB}\rangle\langle w_\alpha^{AB}| \hat{P} \tag{3.8}$$

which leaves us with seemingly two possibilities by which $\hat{\rho}' = |\Phi\rangle\langle\Phi|$. The first possibility is by choosing a projector $\hat{P}_\alpha$ which satisfies

$$\hat{P}_\alpha |w_{\alpha'}^{AB}\rangle = \delta_{\alpha,\alpha'} c |\Phi\rangle. \tag{3.9}$$

with an amplitude $c \in \mathbb{C}$ that obeys $c^* c = p$. We stress that this assumption does not imply $|\Phi\rangle = |w_\alpha^{AB}\rangle$ since the Schmidt vectors with non-zero coefficients do not constitute a *complete* basis for the entire Hilbert space.

A proof by contradiction shows that an apparent second possibility of obtaining a pure state after measurement can be excluded. To this end, assume that there exists a number $n \geq 2$ of Schmidt vectors labelled by the subset of indices

$$\mathcal{Q} \subset \{\alpha | \alpha \in [1, 2, \ldots, \chi]\}$$

and a projection operator $\hat{P}$ for which

$$\hat{P} |w_{\alpha'}^{AB}\rangle = c |\Phi\rangle \quad \forall \quad \alpha' \in \mathcal{Q},$$

where the amplitude $c \in \mathbb{C}$ obeys $c^* c = p/n$. It turns out that this is a conflicting assumption with the properties of a projector and the orthonormality of the Schmidt vectors: For $\alpha', \beta' \in \mathcal{Q}, \alpha' \neq \beta'$ one has, on the one hand,

$$\langle\Phi|\Phi\rangle = \frac{n}{p} \langle w_{\beta'}^{AB} | \hat{P}^2 | w_{\alpha'}^{AB}\rangle = \frac{n}{p} \langle w_{\beta'}^{AB} | \hat{P} | w_{\alpha'}^{AB}\rangle = 1 \tag{3.10}$$

but on the other hand

$$\langle w_{\beta'}^{AB} | w_{\alpha'}^{AB}\rangle = 0 \tag{3.11}$$

which constitutes a contradiction because there exists no projector which satisfies the last equality of (3.10).

Hence, we have singled out the first of these possibilities which leads to a pure state after measurement. The question of whether $|\Phi\rangle$ will be entangled can not be answered in general since this depends on the particular state under consideration. We will discuss a simple example in the following section.

3.3 Example: Supersinglet State of Three Qutrits

A general supersinglet state is defined as [7]

$$|S_N^{(d)}\rangle = \frac{1}{\sqrt{N!}} \sum_{s_{[1,...,N]}} \epsilon_{s_1,s_2,...,s_N} |s_1, s_2, ..., s_N\rangle \tag{3.12}$$

where $s_l = 1, 2, \ldots, d$ and $\epsilon_{s_1,s_2,...,s_N}$ denotes the completely antisymmetric tensor which is zero if two or more indices are equal and assumes $+1(-1)$ for an even (odd) permutation of indices. Supersinglet states arise as ground states of certain permutation Hamiltonians when $d = N$ [8] which shall serve here as an illustration for the concept of projectively extractable entanglement $\mathcal{E}_{PP}$.

Consider the example of a number $N = 3$ so-called qutrit particles which are labelled by A, B, C and whose respective state space is spanned by the vectors $\{|1\rangle, |2\rangle, |3\rangle\}$ and consequently $d = 3$. Suppose these three qutrits are in a supersinglet state

$$|\psi\rangle = |S_3^{(3)}\rangle = \frac{1}{\sqrt{3}} \left(|\psi_{23}^-\rangle_{AB}|1\rangle_C + |\psi_{31}^-\rangle_{AB}|2\rangle_C + |\psi_{12}^-\rangle_{AB}|3\rangle_C \right) \tag{3.13}$$

where we used a shorthand notation for the direct product $|i\rangle \otimes |j\rangle = |i\rangle|j\rangle$ and introduced the generalised singlet state of the particles A and B

$$|\psi_{ij}^-\rangle_{AB} \equiv \frac{1}{\sqrt{2}} (|i\rangle_A|j\rangle_B - |j\rangle_A|i\rangle_B), \quad i, j \in 1, 2, 3.$$

Let us choose a local projector on subsystem A

$$\hat{P}_A = |2\rangle_A\langle 2| + |3\rangle_A\langle 3|$$

and another on subsystem B

$$\hat{P}_B = |2\rangle_B\langle 2| + |3\rangle_B\langle 3|$$

such that $\hat{P} = \hat{P}_A \otimes \hat{P}_B$. The corresponding selective measurement amounts to the contrived scenario of two experimenters, one of which has access to particle A while the other has access to particle B, who conduct simultaneous measurements on their respective particle A and B and independently ask whether their particle is *not* found in state $|1\rangle$ without acquiring further knowledge about the details of the state. One finds that whenever this question can be answered with *yes* by both experimenters the combined system AB is left in the generalised singlet state

$$|\Phi\rangle = |\psi_{23}^-\rangle_{AB}.$$

Since the mentioned measurement outcome is obtained with probability

$$p = \mathrm{Tr}\big[\hat{P}\hat{\rho}_{AB}\big] = \frac{1}{3}$$

and the resulting state has an entropy of entanglement

$$\mathcal{E}(|\Phi\rangle\langle\Phi|) = \ln 2$$

the amount of projectively extractable pure entanglement is

$$\mathcal{E}_{PP}(\hat{\rho}_{AB}) = \frac{1}{3}\ln 2$$

For this example, we chose local projectors of rank two which is equivalent of having a degeneracy in the measurement outcome "both experimenters report *yes*". As stated in the foregoing chapter, a non-degenerate measurement outcome would immediately yield a pure state $|\Phi\rangle$. However, modifying

$$\hat{P}_A = |2\rangle_A\langle 2|, \quad \hat{P}_B = |3\rangle_B\langle 3|$$

gives $|\Phi\rangle = |2\rangle_A|3\rangle_B$ which is separable and hence $\mathcal{E}_{PP}(\hat{\rho}_{AB}) = 0$.

3.4 General Procedure of Projectively Extracting Pure Entangled States

While the choice of the particular measurement in the example of the previous chapter was guided by intuition, often the form of the considered state of the three parties does not easily allow such a guess. We explore in this section the possibility of extracting pure and entangled states from a general state $\hat{\rho}_{AB}$ of the regions to be measured. The general procedure that will be explained below amounts to a systematic algorithm capable of finding the maximal $\mathcal{E}_{PP}$ that can be achieved by local selective measurements.

Recall that under the assumption that the system ABC is in a pure state, the reduced density operator of the considered regions reads as

$$\hat{\rho} = \sum_{\alpha=1}^{\chi} w_\alpha |w_\alpha^{AB}\rangle\langle w_\alpha^{AB}| \qquad (3.14)$$

We may decompose each Schmidt vector using a Schmidt decomposition

$$|w_\alpha^{AB}\rangle = \sum_{\beta=1}^{\chi_\alpha} \sqrt{w_{\alpha,\beta}} |w_{\alpha,\beta}^A\rangle |w_{\alpha,\beta}^B\rangle \qquad (3.15)$$

and construct local projectors of variable rank as follows

$$\hat{P}^A_{\alpha,\mu} = \sum_{\beta \in \mathcal{P}_{\alpha\mu}} |w^A_{\alpha,\beta}\rangle\langle w^A_{\alpha,\beta}|, \qquad (3.16)$$

$$\hat{P}^B_{\alpha,\nu} = \sum_{\beta \in \mathcal{P}_{\alpha\nu}} |w^B_{\alpha,\beta}\rangle\langle w^B_{\alpha,\beta}|. \qquad (3.17)$$

The summations are carried out over subsets of indices $\mathcal{P}_{\alpha\mu}$ and $\mathcal{P}_{\alpha\nu}$ which contain a particular ordered permutations of a number $2 \le r^A_\mu, r^B_\nu \le \chi_\alpha$ of indices. The subscripts $(\alpha\mu)$ label a specific permutation. This leads to local projectors of rank r^A_μ and r^B_ν for region A and B, respectively. A combined projector, concerning both regions, is then constructed as

$$\hat{P}_{\alpha(\mu\nu)} = \hat{P}^A_{\alpha,\mu} \otimes \hat{P}^B_{\alpha,\nu}.$$

For example, for a particular $\mu = 4$ and $\nu = 3$ which label the index permutations $\mathcal{P}_{\alpha 4} = \{2,7\}$ and $\mathcal{P}_{\alpha 3} = \{3,5\}$, hence $r^A_4 = r^B_3 = 2$ and the local projectors would read as

$$\hat{P}^A_{\alpha,4} = |w^A_{\alpha,2}\rangle\langle w^A_{\alpha,2}| + |w^A_{\alpha,7}\rangle\langle w^A_{\alpha,7}| \qquad (3.18)$$

$$\hat{P}^B_{\alpha,3} = |w^B_{\alpha,3}\rangle\langle w^B_{\alpha,3}| + |w^B_{\alpha,5}\rangle\langle w^B_{\alpha,5}| \qquad (3.19)$$

Whether the resultant projector $\hat{P}_{\alpha(\mu\nu)}$ qualifies for the extraction of a pure state can be checked via (3.9). If one or more of the alternative projectors $\hat{P}_{\alpha(\mu\nu)}$ qualify, we proceed by quantifying the entropy of entanglement of the candidate states $|\Phi_{\alpha(\mu\nu)}\rangle$ which are obtained from

$$|\Phi_{\alpha(\mu\nu)}\rangle\langle\Phi_{\alpha(\mu\nu)}| = \mathrm{Tr}\big[\hat{P}_{\alpha(\mu\nu)}\hat{\rho}_{AB}\big]^{-1} \hat{P}_{\alpha(\mu\nu)}\hat{\rho}_{AB}\hat{P}_{\alpha(\mu\nu)}.$$

Finally, we define

$$\mathcal{E}_{PP} = \max_{\alpha(\mu\nu)}\big\{\mathrm{Tr}\big[\hat{P}_{\alpha(\mu\nu)}\hat{\rho}_{AB}\big]\mathcal{E}(|\Phi_{\alpha(\mu\nu)}\rangle\langle\Phi_{\alpha(\mu\nu)}|)\big\} \qquad (3.20)$$

which is the product of the entropy of entanglement of the post-measurement pure candidate state $|\Phi_{\alpha(\mu\nu)}\rangle$ and the corresponding probability of preparing that state, maximised among all candidates.

Since we argued that the set of qualifying $\hat{P}_{\alpha(\mu\nu)}$ corresponds to the only possible selective measurements by which pure states can be extracted, we conclude that $\mathcal{E}_{PP}$ as obtained through our general procedure will quantify the *optimal* performance of extracting pure state entanglement through local projections.

We would like to test this procedure by looking at the ground state of a simple one-dimensional spin model, which will be introduced hereafter.

3.5 The Transverse XY Spin Chain Model

The transverse XY spin chain describes a collection of spin-$\frac{1}{2}$ particles arranged on a one dimensional lattice which is subjected to a uniform magnetic field. On the one hand, two types of nearest neighbour interactions compete among each other, attempting to minimise the energy by aligning spins along the X or Y axes of the coordinate system, respectively. On the other hand, the transverse field—incident in Z direction—tends to align spins perpendicular to the X–Y plane. At zero temperature, the system therefore seeks the optimal balance between these three contradictory orders so as to minimise the energy. It accomplishes this task by means of quantum fluctuations.

The Hamiltonian of the transverse XY model reads as

$$\hat{H} = -\sum_{l=1}^{N}\left[\left(\frac{1+\gamma}{2}\right)\hat{\sigma}_l^x\hat{\sigma}_{l+1}^x + \left(\frac{1-\gamma}{2}\right)\hat{\sigma}_l^y\hat{\sigma}_{l+1}^y + h\hat{\sigma}_l^z\right] \quad (3.21)$$

where $0 \leq \gamma \leq 1$ designates the XY-anisotropy and the parameter $h \geq 0$ governs the transverse magnetic field amplitude. The transverse Ising model corresponds to the limit $\gamma = 1$ while one recovers the transverse XX-model for $\gamma = 0$. We have come across the latter (with zero magnetic field) in the preceding chapter. The thermodynamical properties of the transverse XY model have been exhaustively studied (see [9] and references therein). We illustrate the ground state phase diagram in Fig. 3.2.

Two distinct gapless[1] (quantum critical) regimes in the (h, γ) plane can be identified (Fig. 3.2): The Ising transition corresponds to the line $h = 1$ and $0 < \gamma \leq 1$ while the XX transition occurs at $\gamma = 0$ and $0 \leq h < 1$. We will return to the critical properties of this model in Sect. 4.2 of the following chapter in more detail which will be justified in that context.

For a particular choice of parameters the ground state factorises (becomes separable): This occurs along the disorder line [10–12] $h^2 + \gamma^2 = 1$.

Eigenstates of $\hat{H}$ are also eigenstates of the parity operator $\hat{Z} = \prod_l \hat{\sigma}_l^z$, i.e. $[\hat{H}, \hat{Z}] = 0$. At the isotropic point $\gamma = 0$, the model additionally obeys conservation of total magnetisation along the z-axis, $[\hat{H}, \hat{S}_z] = 0$. At this isotropic point, the ground state z-magnetisation per site varies from $\langle S_z \rangle / N = 0$ for $h = 0$ to $\langle S_z \rangle / N = 1$ for $h = 1$. For finite N, magnetisation does not vary continuously but changes abruptly whenever a level crossing of the ground state energy occurs (see Fig. 3.3 for the example $N = 4$).

In this chapter, exclusively, periodic boundary conditions will be imposed:

$$\hat{\sigma}_{N+1}^\alpha = \hat{\sigma}_1^\alpha, (\alpha = x, y, z).$$

---

[1] The finite system will always display a gap, hence gapless refers to the spectrum of the infinite system.

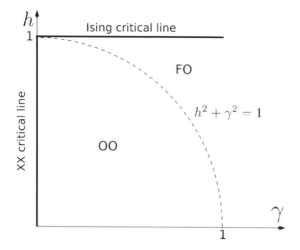

Fig. 3.2 Phase diagram of the transverse XY model. For $h > 1$ the model is paramagnetic. The ground state becomes separable along the disorder line (*dashed curve*), which separates an ordered oscillating phase ($h^2 + \gamma^2 < 1$, "OO") from a ferromagnetic phase ($h^2 + \gamma^2 > 1$, "FO"). The model undergoes quantum phase transitions at $h = 1, 0 < \gamma \leq 1$ and at $\gamma = 0, 0 \leq h < 1$ (*bold lines*)

3.6 Results and Discussion

In the following, we explore the ground state of the transverse XY model for a particular system of $N = 6$ spins arranged on a ring through exact diagonalisation. This system will be partitioned symmetrically into diametrically opposite and

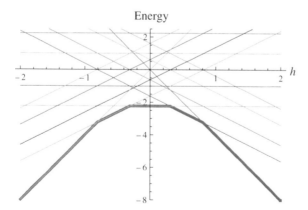

Fig. 3.3 Eigenspectrum of an XX spin chain ($\gamma = 0$) with $N = 4$ sites as a function of the external magnetic field h. The ground state energy is shown as *bold lines*. Upon increasing h, at each level-crossing of the ground state energy the quantum number M increases by one, and the magnetisation displays a discontinuous jump. Reprinted from [13]

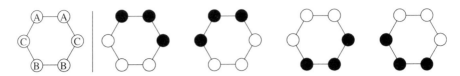

Fig. 3.4 *Left.* Partition of a spin ring with $N = 6$ into the measured regions A and B, and their complement C which gives rise to a spatial separation. *Right.* Four odd parity configurations (*black circles* correspond to $|\uparrow\rangle$ and *white circles* to $|\downarrow\rangle$) of the spin ring, which display individual even parity of the regions A and B and are commensurate with a quantum number $M = 0$ (equal number of up and down spins). Each of these configurations occurs with equal amplitude in the ground state prior to measurement, owing to the discrete lattice symmetry of translation. A local selective parity measurement on A and B (e.g., yielding even parity of both regions) gives rise to a pure and entangled state $|\Phi\rangle$ of the measured regions, see text

contiguous regions A and B, comprising two spins each, that are separated from each other by a single site at either end (Figure 3.4, left).

By numerically applying the general procedure that was explained in Sect. 3.4, we find that nonzero pure entanglement can indeed be extracted by local projections, in specific regions of the ground state phase diagram. Fig. 3.5 shows $\mathcal{E}_{PP}$ as a function of the parameters h and γ.

Figure 3.5 highlights the interplay of having a finite probability of extracting a pure state and the amount of entanglement of this extracted state (top and mid panel, respectively). Only if both conditions are met $\mathcal{E}_{PP}$ assumes a finite value. Considering a slice of the three dimensional graphs at $\gamma = 0$, we see that $\mathcal{E}_{PP}$ is a constant function between two discontinuous jumps at $h \sim \pm 0.5$. This suggests that in this regime of the phase diagram $\mathcal{E}_{PP}$ is linked to the symmetry of conserved z-magnetisation, the corresponding quantum number M of which changes discontinuously at the same values of the external field.

The numerical implementation of the general procedure allows to infer which particular projector has been singled out, and for the regime of nonzero $\mathcal{E}_{PP}$, we find that these projectors correspond to either a local selective measurement of z-magnetisation or local selective parity measurement. While global parity is a conserved quantity for the entire phase diagram, the global z-magnetisation is conserved only at $h = 0$, as pointed out in Sect. 3.5.

Let us have a closer look at the case of a local selective parity measurement. The region of nonzero $\mathcal{E}_{PP}$ for $h \neq 0$ beyond $(h = 0)$, as observed in Fig. 3.5, is numerically found to correspond to a ground state $|\psi\rangle$ with odd parity, that is to say

$$\hat{Z}|\psi\rangle = -|\psi\rangle.$$

The unperturbed ground state must therefore obey certain conditions regarding the local parity of region A, B and C individually: If the local parity of both the regions A and B is even ($|\downarrow\downarrow\rangle_{A(B)}$ or $|\uparrow\uparrow\rangle_{A(B)}$), then the complementary spins must be of odd parity ($|\downarrow\uparrow\rangle_{\overline{AB}}$ or $|\uparrow\downarrow\rangle_{\overline{AB}}$) so as to comply with the global odd parity of ground state (recall that the eigenvalue of $\hat{Z}$ is the product of the parities of region A, B, and their complement).

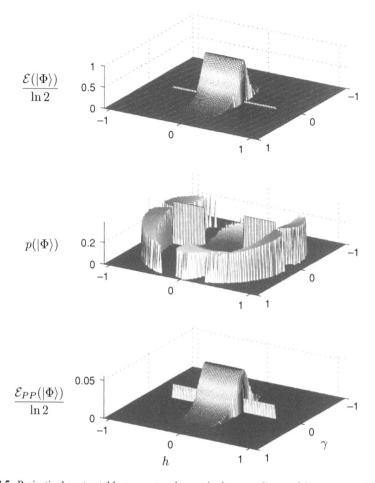

$\dfrac{\mathcal{E}(|\Phi\rangle)}{\ln 2}$

$p(|\Phi\rangle)$

$\dfrac{\mathcal{E}_{PP}(|\Phi\rangle)}{\ln 2}$

h $\qquad\qquad\qquad\qquad$ γ

Fig. 3.5 Projectively extractable pure entanglement in the ground state of the transverse XY spin ring model with $N = 6$. The regions of measurement are diametrically opposite blocks of spins, comprising two spins each. $\mathcal{E}_{PP}$ (*bottom panel*) is obtained from the general procedure, as explained in the text, and amounts to the product of the entropy of entanglement, shown in the top panel, and the probability of preparing a particular pure state by way of a local selective measurement, shown in the mid panel. The factor of $1/\ln 2$ in top and bottom panels is due to the different definition of entropy of entanglement in our paper [3] which assumes a logarithm to the base 2 as opposed to the natural logarithm used in this thesis. Reprinted from [3]

A local measurement of parity which does not discriminate between the particular configurations therefore seems a good candidate for a local selective measurement leading to an entangled pure state, and is indeed singled out by the general procedure as found numerically.

For simplicity, let us restrict ourselves to the region $(h = 0, \gamma = 0)$ where the ground state is characterised by both odd parity and a quantum number $M = 0$, corresponding to an equal number of spin up and spin down. Hence, locally measuring even parity in both regions A and B so that

$$\hat{P}^{A(B)} = |\downarrow\downarrow\rangle_{A(B)}\langle\downarrow\downarrow| + |\uparrow\uparrow\rangle_{A(B)}\langle\uparrow\uparrow|$$

would, under the global constraint of odd parity, be consistent with four possible spin configurations, which occur with equal amplitudes in the ground state by virtue of discrete translational symmetry of the ring (Fig. 3.4, right). Thus after projection, the state of the whole system reads as

$$|\Phi\rangle = \frac{1}{2}(|\uparrow\downarrow\rangle_{\overline{AB}} + |\downarrow\uparrow\rangle_{\overline{AB}}) \otimes (|\uparrow\uparrow\rangle_A|\downarrow\downarrow\rangle_B + |\downarrow\downarrow\rangle_A|\uparrow\uparrow\rangle_B)$$

which was verified numerically in the considered part of the phase diagram. The state of regions A and B is therefore pure and entangled, which is clear by inspection.

We conclude that in the context of many-body ground states, this study sheds light on the interrelation between nonzero $\mathcal{E}_{PP}$ and certain quantum numbers which classify the ground state. In the case of the transverse XY spin ring, these quantum numbers are parity and z-magnetisation. The corresponding observables commute with the underlying Hamiltonian, and admit a meaningful measurement both globally and locally.

References

1. C. Cohen-Tannoudji, B. Diu, F. Laloe, *Quantum Mechanics*. (Wiley-Interscience, New York, 2006)
2. H.-P. Breuer, F. Petruccione, *The Theory of Open Quantum Systems* (Clarendon Press, Oxford, 2002)
3. J. Molina, H. Wichterich, V.E. Korepin, S. Bose, Extraction of pure entangled states from many-body systems by distant local projections, Phys. Rev. A **79**(6), 062310 (2009)
4. F. Verstraete, M. Popp, J.I. Cirac, Entanglement versus correlations in spin systems, Phys. Rev. Lett. **92**(2), 027901 (2004)
5. T.J. Osborne, M.A. Nielsen, Entanglement in a simple quantum phase transition. Phys. Rev. A **66**(3), 032110 (2002)
6. A. Osterloh, L. Amico, G. Falci, R. Fazio, Scaling of entanglement close to a quantum phase transition. Nature **416**, 608–610 (2002)
7. A. Cabello, Supersinglets. J. Mod. Opt. **50**, 10049 (2003)
8. C. Hadley, S. Bose, Multilevel multiparty singlets as ground states and their role in entanglement distribution. Phys. Rev. A **77**(5), 050308 (2008)
9. M. Henkel, *Conformal Invariance and Critical Phenomena* (Springer, Berlin, 1999)
10. I. Peschel, V. Emery, Z. Phys. B **43**, 241 (1981)
11. J. Kurmann, H. Thomas, G. Mueller. Phys. A **112**, 235 (1982)
12. S.M. Giampaolo, G. Adesso, F. Illuminati, Theory of ground state factorization in quantum cooperative systems. Phys. Rev. Lett. **100**(19), 197201 (2008)
13. W. Son, V. Vedral, On the quantum criticality in the ground and the thermal states of xx model. OSID **2–3**, 16 (2009)

Chapter 4
Scaling of Negativity Between Separated Blocks in Spin Chains at Criticality

In this chapter we quantify the entanglement between separated blocks in spin chain models and study its behaviour in the vicinity of quantum critical points in terms of negativity. The numerics suggest that at the transition negativity is *scale invariant* in that it is a function of the ratio of the separation to the length of the blocks. We observe that this entanglement displays an exponential decay for large separations of the block and therefore markedly differs from behaviour of correlation functions which at criticality decay according to a power law. We study universal features of negativity and show that it is largely independent of microscopic system parameters for models which belong to the same universality class.

4.1 Introduction

In the past decade a vast interest in entanglement in ground states of many body systems has emerged and a considerable body of results has been obtained [1]. At quantum phase transitions, i.e. a situation where at zero temperature correlations become particularly pronounced, entanglement displays several interesting features. Figure 4.1 illustrates different forms of entanglement that can be studied in this context.

The study of phase transitions has a long standing tradition in condensed matter theory [2]. Conventionally, transitions can occur between gaseous, fluid and solid states upon varying pressure or temperature. An order parameter differs markedly on each side of the transition, such as the particle density in the mentioned example. At zero temperature, a transition between different phases of matter can be triggered by microscopic coupling parameters or external electric or magnetic fields. Quantum fluctuations between qualitatively different ground states dominate the physics on the macroscopic scale, giving rise to the term *quantum phase transition* [3].

H.C. Wichterich, *Entanglement Between Noncomplementary Parts of Many-Body Systems*, Springer Theses, DOI: 10.1007/978-3-642-19342-2_4,
© Springer-Verlag Berlin Heidelberg 2011

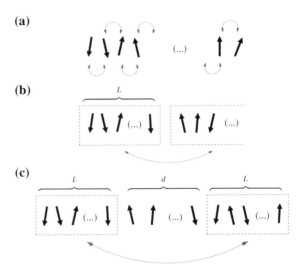

Fig. 4.1 Forms of entanglement. *Top panel* **a**. Entanglement between individual spins is usually rather short ranged, even at quantum critical points where classical correlations exist on all length scales [4, 5]. *Mid panel* **b**. Entanglement between a contiguous block of spins of length L, and the remaining spins. At quantum critical points, entropy of entanglement shows universal scaling behaviour $\mathcal{E} \sim \frac{c}{6} \ln L$ [6–8], where c is the conformal anomaly number (or central charge), defining the universality class of the corresponding continuum model that the model can be associated with at a quantum phase transition [9]. *Bottom panel* **c** . Entanglement between disjoint blocks of spins of length L, separated by a number d of spins. This form will be under investigation in this chapter. Reprinted from [10]

A phase transition is a distinguished example of an emergent phenomenon, i.e. a collective effect that arises in large samples of correlated particles. Usually, thermodynamic properties display generic behaviour that is largely independent of the details of the interaction among constituent particles as well as other characteristics which would matter on the microscopic scale. This feature is called *universality*, and has also been seen to have a bearing on entanglement at quantum phase transitions [5, 6]. One of the most prominent examples in this context is the finding that in the case of ground states of quantum critical spin-1/2 chains entropy of entanglement of a contiguous block of length L displays a leading divergent term [6–8]

$$\mathcal{E} \sim v \frac{c}{6} \ln L \tag{4.1}$$

where c denotes the so called central charge, which is characteristic of the underlying quantum phase transition for a whole host of different microscopic models [9]. The number $v = 1, 2$ corresponds to the number of connections, that the contiguous block has with the rest of the chain. For example, in quantum chains with linear dispersion relation near the Fermi points $\epsilon_k \sim v_F |k|$ the central charge shows up in the low temperature limit $T \ll 1$ of the specific heat [11, 12]

$$C = \frac{\pi c}{3} \frac{k_B^2}{\hbar v_F} T$$

where we temporarily restored all constants. The Ising quantum phase transition of the transverse XY model is characterised by $c = \frac{1}{2}$ whereas the XX transition displays a central charge $c = 1$.

Moreover, the scaling of entanglement of a distinguished region with its extent L has an important implication for the simulability of quantum mechanical states using classical computers [13], and we will comment on this relationship in the context of a simulation algorithm in Sect. 4.3.

Entanglement between individual spins is of interest from a quantum communication perspective. We argued in Chap. 2 that this form of entanglement is notoriously short ranged, even at quantum critical points. The entanglement between neighbouring spins as measured by the concurrence [14] displays several universal scaling features [4, 5]. Furthermore, concurrence heralds a transition in finite size systems by developing singularities in its derivative with respect to the coupling parameter driving the transition.

We unveil in this chapter, that in the case of spin chains the respective properties of these two forms of entanglement can be understood from a common perspective: In fact, one identifies their qualitative properties as limiting cases of the universal behaviour of entanglement between separated blocks of spins. The transverse XY spin chain (Sect. 3.5) will be the model under consideration in this chapter. In Sect. 4.2 we start with a brief presentation of the quantum phase transitions that are displayed by this model. After giving a motivation for our chosen methodology, we proceed with explaining the technical steps that are needed to arrive at the negativity of separated blocks of large blocks of spins in Sect. 4.3. We finally present and discuss our results on universality and scaling of negativity in Sects. 4.4 and 4.5.

The work in this chapter has been done in collaboration with Dr. Javier Molina (Universidad de Cartagena, Spain), and with my supervisor Prof. Sougato Bose.

4.2 The Quantum Phase Transitions of the Transverse XY Spin Chain

We introduced the transverse XY spin chain in Sect. 3.5. For convenience, we recall that the Hamiltonian reads as

$$\hat{H} = -\sum_{l=1}^{N} \left[\left(\frac{1+\gamma}{2} \right) \hat{\sigma}_l^x \hat{\sigma}_{l+1}^x + \left(\frac{1-\gamma}{2} \right) \hat{\sigma}_l^y \hat{\sigma}_{l+1}^y + h\hat{\sigma}_l^z \right]. \qquad (4.2)$$

Here and in the following, open boundary conditions will be imposed:

$$\hat{\sigma}_{N+1}^\alpha = 0, \quad \alpha = x, y, z$$

A quantum critical point is associated with a continuum of gapless excitations which governs the low energy physics [10]. Two such regimes can be identified in the model under consideration, the XX critical line and the Ising critical line, as pointed out earlier (Fig. 3.2). In this section we will have a closer look at the single particle energy spectra, which reveal these two distinct critical phases.

The model can be solved by mapping it to spinless fermions [9], along essentially the same lines which were presented for the simple case of the XX model without external field (Appendix A). This procedure gives rise to a free fermion Hamiltonian which assumes, up to constant terms,

$$\hat{H} \sim \sum_k \epsilon_k \hat{\eta}_k^\dagger \hat{\eta}_k \qquad (4.3)$$

with single particle energy spectrum [9]

$$\epsilon_k = 2\sqrt{(h - \cos q_k)^2 + \gamma^2 \sin^2 q_k}, \qquad (4.4)$$

where the momenta $0 \le q_k < 2\pi$, $k = 0, 1, \ldots, N - 1$ slightly depend on the chosen boundary conditions. Quite generally $|q_{k+1} - q_k| \sim N^{-1}$ [9], hence as N becomes large q_k approaches a continuous function q. The critical properties are revealed in the limit of vanishing momentum of the single particle spectrum (this corresponds to the low-lying excitations when $h \ge 1$)

$$\epsilon_k\big|_{q \to 0} = 2\sqrt{(h - 1)^2 + ((h - 1) + \gamma^2)q^2 + \mathcal{O}(q^4)}. \qquad (4.5)$$

One recognises that in the thermodynamic limit and along the Ising critical line $0 < \gamma \le 1$ (see Fig. 3.2) the gap vanishes as $\sim |h - 1|$. Regarding the isotropic point $\gamma = 0$, it is seen from (4.4) that the gap vanishes for $q = \arccos(h), h < 1$, which happens to coincide with the Fermi wave number [15]. Hence, we have identified the two zero temperature critical phases which are displayed by the transverse XY spin chain.

4.3 Method of Quantifying Entanglement Between Disjoint Regions in Spin Chains

In contrast to the study of entanglement of a contiguous block of spins with the remainder of the spin chain, which can be largely handled analytically for the models at hand, rather substantial obstacles arise when attempting to quantify entanglement between disjoint regions of spins in spin chains. This seems surprising at first sight, given that the model under consideration is commonly regarded as being completely solvable.

On the one hand, the reduced density operator (RDO) of disjoint blocks of spins assumes no simple structural expressions for the ground state of the chain, in general, even if it maps to a free fermion theory [16, 17]. This stands in sharp contrast to the RDO of a contiguous block of length L where it is a simple exponential $\sim e^{-H}$ involving a quadratic form H of Jordan–Wigner fermion operators [6, 18]. Entanglement entropy can then be obtained by diagonalising an $L \times L$ dimensional matrix which, in simple cases, is given by the collection of two point correlation functions $C_{i,j} = \langle \hat{c}_i^\dagger \hat{c}_j \rangle$ [6, 18]. We will exploit this method to study the limit of adjacent blocks in Sect. 4.5.

On the other hand, even if it were possible to derive such an equivalent expression in terms of fermionic operators one would have no immediate advantage from it: We argued that the quantification of entanglement *between* disjoint regions A and B requires measures such as negativity. Presently, the literature is lacking a useful result that would enable the computation of negativity based on the fermionic correlation functions. Consequently, one would have to numerically construct the full density operator in the computational basis which is feasible only for small system sizes due to the enormous growth of matrix dimensions. Moreover, the full matrix could then be obtained, with less effort, by exact diagonalisation in the original spin representation.

This state of affairs suggests a numerical approach, and we will explain in the following how to apply the concepts of matrix-product-states (MPS) or density matrix renormalisation group (DMRG) to the problem at hand. At the heart of MPS and DMRG in one dimensional composite quantum systems lies the Schmidt decomposition (SD) (see (Sect. 1.1.2). Consider a quantum state defined on a product Hilbert space $\mathcal{H} = \mathcal{H}_1 \otimes \mathcal{H}_2 \otimes, \ldots, \mathcal{H}_N$ of individual Hilbert spaces $\mathcal{H}_l$ of spin-1/2 particles. In the following, let $\mathcal{H}_l (l = 1, \ldots, N)$ be spanned by the $\hat{\sigma}_l^z$ eigenstates $|\sigma_l\rangle \in \{|\uparrow\rangle, |\downarrow\rangle\}$. A SD is defined for each bond l which partitions the chain into left and right parts giving rise to an overall $N - 1$ possible SD's. The finite system DMRG algorithm which was devised in [19] amounts to optimally approximating (in a sense to be made precise below) all these SD's along with the basis transformations which relate an SD at bond l to that of $l - 1$ and $l + 1$.

Let us start the discussion with the SD at bond 1, thereby conceptually following [20],

$$|\psi\rangle = \sum_{\alpha_1} \sqrt{w_{\alpha_1}} |w_{\alpha_1}^L\rangle \otimes |w_{\alpha_1}^R\rangle, \quad \alpha_1 = 1, \ldots, \chi_1. \tag{4.6}$$

Recall that the Schmidt rank χ_l is always bounded by the smaller of the two Hilbert space dimensions of $\mathcal{H}^L = \mathcal{H}_1 \otimes \cdots \otimes H_l$ and $\mathcal{H}^R = \mathcal{H}_{l+1} \otimes \cdots \otimes H_N$, hence $\chi_1 \leq 2$.

The SD of the neighbouring bond $l = 2$ is obtained from (4.6) using the following definitions

$$|w_{\alpha_1}^R\rangle \equiv \sum_{\alpha_2} \sum_{\sigma_2} \Lambda_{\alpha_1,\alpha_2}^{[2]\sigma_2} \sqrt{w_{\alpha_2}} |\sigma_2\rangle \otimes |w_{\alpha_2}^R\rangle, \quad \alpha_2 = 1, \ldots, \chi_2 \tag{4.7}$$

inserting in (4.6) gives

$$|\psi\rangle = \sum_{\alpha_1,\alpha_2} \sum_{\sigma_2} \sqrt{w_{\alpha_1}} \Lambda^{[2]\sigma_2}_{\alpha_1,\alpha_2} \sqrt{w_{\alpha_2}} |w^L_{\alpha_1}\rangle \otimes |\sigma_2\rangle \otimes |w^R_{\alpha_2}\rangle, \qquad (4.8)$$

$$|w^L_{\alpha_2}\rangle \equiv \sum_{\alpha_1} \sum_{\sigma_2} \sqrt{w_{\alpha_1}} \Lambda^{[2]\sigma_2}_{\alpha_1,\alpha_2} |w^L_{\alpha_1}\rangle \otimes |\sigma_2\rangle, \qquad (4.9)$$

$$\Leftrightarrow \quad |\psi\rangle = \sum_{\alpha_2} \sqrt{w_{\alpha_2}} |w^L_{\alpha_2}\rangle \otimes |w^R_{\alpha_2}\rangle. \qquad (4.10)$$

Clearly, we can iterate this procedure until we reach the rightmost bond at $l = N - 1$. The boundaries require a slight modification

$$|w^L_{\alpha_1}\rangle = \sum_{\sigma_1} \Lambda^{[1]\sigma_1}_{\alpha_1} |\sigma_1\rangle, \qquad (4.11)$$

$$|w^R_{\alpha_{N-1}}\rangle = \sum_{\sigma_N} \Lambda^{[N]\sigma_N}_{\alpha_{N-1}} |\sigma_N\rangle. \qquad (4.12)$$

In the computational basis the state therefore decomposes into

$$|\psi\rangle = \sum_{\sigma_{[1,\dots,N]}} \left(\sum_{\alpha_{[1,\dots,N-1]}} \Lambda^{[1]\sigma_1}_{\alpha_1} \sqrt{w_{\alpha_1}} \Lambda^{[2]\sigma_2}_{\alpha_1\alpha_2} \dots \sqrt{w_{\alpha_{N-1}}} \Lambda^{[N]\sigma_N}_{\alpha_{N-1}} \right) |\sigma_1, \sigma_2, \dots, \sigma_N\rangle \quad (4.13)$$

which suggests that each coefficient can be interpreted as a matrix product that is terminated to the left and right by vectors so as to give rise to a scalar. We are left with a compact matrix product representation of an arbitrary state on $\mathcal{H}$

$$|\psi\rangle = \sum_{\sigma_{[1,\dots,N]}} A^{[1]\sigma_1} A^{[2]\sigma_2} \dots A^{[N]\sigma_N} |\sigma_1, \sigma_2, \dots, \sigma_N\rangle \qquad (4.14)$$

where we introduced matrices (and boundary vectors)

$$\left(A^{[1]\sigma_1}\right)_{\alpha_1} \equiv \Lambda^{[1]\sigma_1}_{\alpha_1} \sqrt{w_{\alpha_1}} \qquad (4.15)$$

$$\left(A^{[l]\sigma_l}\right)_{\alpha_{l-1},\alpha_l} \equiv \Lambda^{[l]\sigma_l}_{\alpha_{l-1}\alpha_l} \sqrt{w_{\alpha_l}} \quad (l = 2, \dots, N - 1) \qquad (4.16)$$

$$\left(A^{[N]\sigma_N}\right)_{\alpha_{N-1}} \equiv \Lambda^{[N]\sigma_N}_{\alpha_{N-1}}. \qquad (4.17)$$

At this point, it will be useful to introduce a pictorial description in order to elucidate how our study of entanglement between separated blocks is carried out. We started above with a representation of the ground state in terms of the SD at bond $l = 1$. We may represent this setting pictorially as follows

$$[L_1][R_{N-1}],$$

where square brackets designate a Schmidt basis or *block* representation, "*L, R*" label left and right blocks, and the subscript indices represent the number of spins in the blocks respectively.

In the second step we represented the SD at bond $l = 2$ "$[L_2][R_{N-2}]$" in terms of its preceding SD at $l = 1$. The computational steps carried out in going from (4.6) to (4.10) can then be depicted as follows:

$$[L_1][R_{N-1}] \rightarrow [L_1] \bullet [R_{N-2}] \rightarrow [L_2][R_{N-2}],$$

where the bold dot represents the computational basis of a single spin.

The tensors Λ and $\sqrt{w}$ which occur in (4.13) are obtained through a variational procedure called DMRG, the details of which are explained in Appendix D. In practise, a DMRG calculation will not retain all Schmidt basis vectors for a given bipartition. In fact, one truncates the SD—ordered term-wise with decreasing weights $\sqrt{w_\alpha}$—at the Mth term. The number M is called *bond-dimension*. As a result, the tensor dimensions of $\Lambda^{[l]\sigma_l}_{\alpha_{l-1}\alpha_l}$ and $\sqrt{w_{\alpha_l}}$ are all bounded by M. A figure of merit which is commonly used to judge the accuracy of the DMRG method is the so-called truncated weight

$$\epsilon = \sum_{\alpha > M} w_\alpha. \qquad (4.18)$$

In this sense, DMRG gives rise to an optimal approximation to the actual ground state coefficients for a given bond dimension M.

A complication arises for quantum critical systems, where the entropy of entanglement diverges logarithmically with the size of a block (4.1). In the case of the left block in DMRG, one has a single boundary of the block with the rest of the chain, which gives rise to the entropy scaling $(1 \ll l \ll N)$ [7]

$$\mathcal{E} = -\sum_{\alpha_l=1}^{\chi_l} w_{\alpha_l} \ln w_{\alpha_l} \sim \frac{c}{6} \ln l. \qquad (4.19)$$

A rough estimate on the scaling of the Schmidt rank can be obtained by assuming that all Schmidt values are equal (for a given $\mathcal{E}$ this would yield a minimal χ_l). Hence, one obtains that the actual Schmidt rank diverges at least as a power of the subsystem size

$$\chi_l \sim l^{c/6} \qquad (4.20)$$

which inhibits the study of arbitrarily large system sizes N with DMRG at criticality.

Let us now turn to the quantification of entanglement between disjoint blocks of spins. The starting point is the SD of a symmetric bisection of a spin chain (assuming even N)

$$[L_{N/2}][R_{N/2}].$$

Once all matrices $A^{[l]}$ are obtained through DMRG, we are in a position to change to a representation

$$[L_{N/2-1}] \bullet \bullet [R_{N/2-1}].$$

The next step amounts to forming the density operator in this representation $\hat{\rho} = |\psi\rangle\langle\psi|$ and trace out those degrees of freedom which are associated to the separating spins "$\bullet\bullet$". This leaves us with the RDO $\hat{\rho}_{AB}$ in the representation

$$[L_{N/2-1}][R_{N/2-1}]$$

corresponding to the state of two blocks of spins separated by $d = 2$ spins. This procedure is iterated, yielding a set of operators $\hat{\rho}_{AB}$ for increasing separations $d = 2, 4, 6, \ldots$.[1] It follows, that one can perform the otherwise computationally demanding steps of partial transposition

$$(\hat{\rho}_{AB})_{\alpha,\beta,\gamma,\delta} = \left(\hat{\rho}_{AB}^{T_B}\right)_{\gamma,\beta,\alpha,\delta} \tag{4.21}$$

and subsequent diagonalisation of $\hat{\rho}_{AB}^{T_B}$ with comparative ease in the compact representation of Schmidt bases: By way of the procedure explained above, $\hat{\rho}_{AB}$ is represented by a matrix with entries

$$(\hat{\rho}_{AB})_{\alpha,\beta,\gamma,\delta} = \langle w_{\alpha_{(N-d)/2}}^L | \otimes \langle w_{\beta_{(N+d)/2}}^R | \hat{\rho}_{AB} | w_{\gamma_{(N-d)/2}}^L \rangle \otimes | w_{\delta_{(N+d)/2}}^R \rangle \tag{4.22}$$

and with matrix dimensions of at most $M^2 \times M^2$. The negativity $\mathcal{N}$ is obtained from the eigenvalues of $\hat{\rho}_{AB}^{T_B}$, and (1.20). This is the computational step which, together with the individual amount of available random access memory (RAM), limits the value of the bond dimension M. With 2 Gb of RAM, we find that $M \sim 60$ sets the upper limit.

4.4 Negativity of Disjoint Blocks of Spins

In a first step, we study negativity in the vicinity of the Ising transition at a critical value $h_c = 1$ of the parameter h of the transverse magnetic field. In Fig. 4.2 we show for different system sizes N and a fixed size ratio $\mu = d/L = 1/3$ the negativity as a function of the deviation $h - h_c$. While the size dependence of negativity is rather pronounced in the non-critical regime, at exactly $h = h_c$ all curves for different N coincide at a certain value of negativity. This is the main insight from this first study. Scale invariance is an ubiquitous feature of systems at criticality [2, 21]. From a broader perspective, scale invariance is a signature of a more general feature of systems at criticality, namely that of *scaling*. In its

---

[1] odd values of d could be obtained with only slight modifications, but we restrict ourselves to symmetric partitions and even values of d in this treatise.

simplest form, this means that two measurable quantities depend on each other in a power-law fashion. For instance, a thermodynamic quantity $\phi(x)$ which is supposed to be a function of the spatial coordinate x (for dimensioned quantities) would usually behave under a scale transformation $x \to bx$ as

$$\phi(bx) \sim b^{y_\phi} \phi(x) \tag{4.23}$$

where y_ϕ is a characteristic exponent and where we omitted constants which would make the relation dimensionally consistent. Scaling laws like (4.23) herald a power-law spatial dependence, and we see that *qualitatively* the physics would not depend on the scale one uses to observe it. This could already be the definition of scale invariance, but a concise definition usually requires the notions of renormalisation group [21]. We only intend to give a flavour of the canonical notion of scale invariance here, so as to underline that the variant of scale invariance exhibited by negativity is indeed different and *manifest*: Negativity is invariant with respect to increasing the sizes of the regions and that of the separation by a common factor (in fact, this is *not* a scale transformation in the aforementioned sense).

It is questionable that a simple dimensional analysis would lead to the same conclusion. Since negativity is a measure of correlations it is always dimensionless. If all dimensions in our study were restored, it would depend on the dimensioned quantities $d' = da$ and $L' = La$, where a is the unit of the interparticle distance. Dimensionally consistent dependencies would therefore be $\mathcal{N}(d'/L')$ which is scale invariant in the manifest sense or $\mathcal{N}(d'/a, L'/a)$ which is not.

A similar set of graphs as in Fig. 4.2 can be plotted for other values of the anisotropy $0 \leq \gamma < 1$, and while one obtains qualitatively the same graph with different magnitudes in the proximity of the transition, exactly at the critical point negativity turns out to be independent of γ. We discuss this feature more closely in the second half of this section.

Let us summarise the two important features of negativity that are predicted by the first part of this numerical study. At the quantum critical point

- Negativity of disjoint blocks is manifestly scale invariant and
- Independent of the model parameter γ in the universality class of the Ising transition.

We will now investigate negativity at the critical points ($h = 1, 0 < \gamma \leq 1$) and ($0 \leq h < 1, \gamma = 0$) more closely. If upon doubling the system size N at the critical point we detect no discernible change in the value of $\mathcal{N}$ with an accuracy of 3 (5) significant figures in the XX (Ising) critical regime, we will regard negativity as being converged. In the Ising case, convergence sets in upon doubling the system size from $N \sim 256$ to ~ 512 in the case of the Ising phase transition. We also note that the bond dimension M is varied with the system size, so as to fix the truncated weight $\epsilon \sim 10^{-10}$. Simulation of the XX critical phase is harder regarding the DMRG simulation [here, $c = 1$ compared to $c = 1/2$ in the Ising critical regime

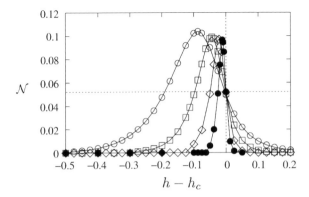

Fig. 4.2 Scale invariance of negativity at the critical point of the transverse Ising model ($\gamma = 1$ in (3.21)). The curves represent data for different system sizes $N = 32$ (*open circles*), $N = 64$ (*squares*), $N = 128$ (*diamonds*), and $N = 256$ (*filled circles*). The ratio of (block separation):(block length) is fixed to $\mu = d/L = 2/3$. The quantum phase transition occurs at $h = h_c = 1$ where all curves coincide beyond $N \sim 256$ within the numerically achievable accuracy. This scale invariant point lies at $(h - h_c = 0, \mathcal{N} = 0.052)$ for this choice of μ and is highlighted with the *dashed lines*. The finite-size shift of the peak value of negativity vanishes as a power law $\sim N^{-1}$, which is often observed for physical quantities near the transition, heralding the presence of a critical point [21]. However, the height of the peak of negativity decreases with N, suggesting that there will be no singularity at $h = h_c = 1$ and $N \to \infty$. Figure reprinted from [10]

which leads to a stronger divergence of the Schmidt rank, see (4.20)], but it is seen that convergence sets in earlier. Hence, we will conduct our numerical study for system sizes of $N = 96$ and $N = 256$ for XX and Ising critical points, respectively, where the higher numerical accuracy can be achieved for the latter.

Negativity is then plotted as a function of the ratio $\mu = d/L$, for different values of the anisotropy γ, see Fig. 4.3. As stated above, we find that negativity is independent of the anisotropy when varying it within the parameter regime $0 \le \gamma < 1$, which defines the Ising critical transition line. At the second phase transition, which occurs at the isotropic XY point $\gamma = 0$ and for the magnetic field varying within $0 \le |h| < 1$, we see a quantitative difference in $\mathcal{N}$, commensurate with the different universality class that is associated with this transition.

For both the Ising and the XX transition we observe a common decay behaviour. In the semi-logarithmic scale of the graph in Fig. 4.3, one recognises a crossover between what seems like a power-law decay for $\mu < 1$ (corresponding to small separation and large blocks) and an exponential decay for $\mu > 1$. One could, on the one hand, attribute this behaviour to the fact that the sizes of the blocks become comparable to the overall system-size $N = 2L + d$. On the other hand, qualitatively similar results have been obtained for the continuum limit of chains of harmonic oscillators, where the blocks are pieces of an infinitely extended system [22]. We therefore expect that in spin chains of infinite extent, one would observe a similar crossover phenomenon for $\mu \sim 1$.

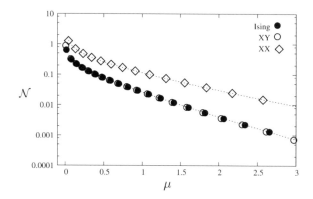

Fig. 4.3 Negativity as a function of the ratio $\mu = d/L$ in a semi-logarithmic scale. The data shown correspond to the two distinct critical regimes which are displayed by the transverse XY model [see (3.21)]. The two data sets within the Ising critical regime, i.e. Ising ($h = 1, \gamma = 1$) and XY ($h = 1, \gamma = 1/2$) fall onto the same curve, while the data for the XX model at vanishing field ($h = 0, \gamma = 0$) quantitatively differ. This is commensurate with the different universality classes to which these models belong. The decay of negativity displays a characteristic crossover between what seems to be a power-law decay for ($\mu < 1$) and an exponential decay for large separations ($\mu > 1$). We model the data with a functional ansatz of the form $\mathcal{N} \sim \mu^{\beta} e^{-\alpha \mu}$, which yields an accurate fit to our data (*dotted lines*). Reprinted from [10]

By making an ansatz

$$\mathcal{N} \sim \mu^{-\beta} e^{-\alpha \mu} \tag{4.24}$$

for the functional form of negativity, allowing for adjustable parameters α and β, one can achieve an accurate least-squares fit to the data points of XX and Ising critical regimes respectively (see Fig. 4.3). The precise values of α and β are slightly sensitive (in the second decimal place) to the chosen fitting interval. By fixing the latter to $0.1 \leq \mu \leq 3$ we find for the two transitions

$$\text{Ising:} \quad \alpha = 1.68, \quad \beta = 0.38, \tag{4.25}$$

$$\text{XX:} \quad \alpha = 0.96, \quad \beta = 0.47, \tag{4.26}$$

which might serve as a hint for future attempts to relate these values to the decay exponents of correlation functions, for instance. In the case of the XX model, it is known [23] that the dominant correlation functions asymptotically decay as $\langle \hat{\sigma}_i^x \hat{\sigma}_j^x \rangle \sim \langle \hat{\sigma}_i^y \hat{\sigma}_j^y \rangle \sim |i - j|^{-1/2}$, while the remaining correlation function $\langle \hat{\sigma}_i^z \hat{\sigma}_j^z \rangle$ decays far quicker, namely as $\sim |i - j|^{-2}$. On these grounds, and by comparing the fitted value $\beta = 0.47$ one might expect that in the case of the XX critical regime β is linked to the decay exponent $1/2$ of the dominant correlation functions. A cross check of this behaviour in the Ising critical regime, however, does not support this conjecture. In the subsequent section we study negativity in the limit of adjacent blocks by analytical means, which verifies the polynomial divergence as $\mu \to 0$

and suggests that α could instead be related to the central charge c, albeit no conclusive link has been established, yet.

The emergence of a *dimensionless scale*, given by the exponent α, is rather surprising in that correlations are usually expected to decay in a power-law fashion at a quantum critical point [3]. Despite extended efforts, the determined values of α and β could not be unequivocally related to known characteristic exponents of the underlying models as yet. Qualitatively, one can attribute this exponential decay with a property of entanglement called *monogamy*: Each spin is seen to have a finite capacity to entangle with other spins. In contrast, the amplitude of classical correlations is not restricted in a setting with multiple parties and can be maximal among all of them, in principle. If among nearest neighbours this capacity of entanglement is almost exhausted (this was evidenced in [4, 5]), clearly entanglement must become strongly suppressed for large separations.

Finally, from a quantum information perspective, it is an interesting observation that while pairs of spins are usually unentangled beyond a separation of few lattice sites, blocks of spins can become entangled across substantial distances, as long as their respective extent is sufficiently large. This behaviour was therefore attributed to multipartite entanglement in [1].

4.5 Negativity in the Limit of Adjacent Blocks of Spins

The problem of quantifying negativity in the case of adjacent blocks becomes comparatively simple, and can be studied on the basis of the exact solution of the transverse XY spin chain. We do so in this section for the case $\gamma = 0$ and $h = 0$ which corresponds to the XX spin chain (Sect. 2.1). The diagonalisation of this model is presented in Appendix A. Note that the present definition of the Hamiltonian differs by an overall minus-sign and $J = 1$ in comparison to the XX model of Chap. 2 and Appendix A which can be absorbed into the single particle energies

$$\epsilon_k = -2\cos\left(\frac{\pi k}{N+1}\right) \tag{4.27}$$

so that the ground state corresponds to an occupation of all fermionic modes with $k \leq \frac{N+1}{2}$ (Fermi sea).

Since the state of the two blocks is pure, in this case negativity can be obtained by formula [see (1.26)]

$$\mathcal{N} = \left(\text{Tr}\left[\sqrt{\hat{\rho}_L}\right]\right)^2 - 1$$

where $\hat{\rho}_L$ denotes the reduced density operator of the right or left half of the chain (these operators are identical for chains of an even number of sites due to reflection symmetry about the middle of the chain), which are both of length L. It is shown in Ref. [18] that the RDO assumes the form

$$\hat{\rho}_L = Ke^{-\tilde{\hat{H}}}$$

where K ensures normalisation and where the exponent can be represented by a free fermion operator

$$\tilde{\hat{H}} = \sum_k \tilde{\epsilon}_k \hat{f}_k^\dagger \hat{f}_k.$$

Hence, the matrix representation of the density operator of the block factorises with respect to the multi-mode Fock basis of the fermion operators $\hat{f}_k$

$$\rho_L = K\rho_1 \otimes \rho_2 \otimes \cdots \otimes \rho_L$$

with single particle density matrices (unnormalised)

$$\rho_k = \begin{pmatrix} 1 & 0 \\ 0 & e^{-\tilde{\epsilon}_k} \end{pmatrix}.$$

Using $\mathrm{Tr}[\rho_1 \otimes \rho_2 \otimes \cdots \otimes \rho_L] = \prod_k \mathrm{Tr}[\rho_k]$ gives for the normalisation constant

$$\mathrm{Tr}[\rho_L] = 1 = K \prod_{k=1}^{L}(1 + e^{-\tilde{\epsilon}_k}) \Leftrightarrow K = \prod_{k=1}^{L} \frac{1}{1 + e^{-\tilde{\epsilon}_k}}. \tag{4.28}$$

The pseudo-energies $\tilde{\epsilon}_k$ are related to the eigenvalues ζ_k of the correlation matrix $C_{i,j} = \langle \hat{c}_i^\dagger \hat{c}_j \rangle$, where the $\hat{c}_i$ denote the Jordan–Wigner fermion operators (2.19). One obtains (see (B.21), and [18])

$$\tilde{\epsilon}_k = \ln \frac{1 - \zeta_k}{\zeta_k} \Leftrightarrow K = \prod_{k=1}^{L}(1 - \zeta_k) \tag{4.29}$$

For negativity, we find

$$\mathcal{N} = \left(\mathrm{Tr}\left[\sqrt{\hat{\rho}_L}\right]\right)^2 - 1 = K\left[\prod_{k=1}^{L}(1 + e^{-\tilde{\epsilon}_k/2})\right]^2 - 1 \tag{4.30}$$

$$\mathcal{N} = \left[\prod_{k=1}^{L}\sqrt{(1 - \zeta_k)}\left(1 + \sqrt{\frac{\zeta_k}{1 - \zeta_k}}\right)\right]^2 - 1 \tag{4.31}$$

$$\mathcal{N} = \left[\prod_{k=1}^{L}\left(\sqrt{1 - \zeta_k} + \sqrt{\zeta_k}\right)\right]^2 - 1 \tag{4.32}$$

The correlation matrix is known analytically in terms of the coefficients $g_{k,l}$ (2.28)

$$C_{i,j} = \langle \hat{c}_i^\dagger \hat{c}_j \rangle = \sum_{m=1}^{L} g_{i,m} g_{m,j}.$$

Here, index m runs over values for which the single particle energies of the system are negative (this corresponds to the Fermi sea). By numerically diagonalising C, we obtain the eigenvalues ζ_k which subsequently allows to compute $\mathcal{N}$.

A prediction from conformal field theory for negativity can be given in this limit, and for large L, based on the result from [7]

$$\mathrm{Tr}[\hat{\rho}_L^n] \sim L^{-(c/12)(n-1/n)} \tag{4.33}$$

which holds for a finite system of length N with open boundaries, partitioned at an interior point into left and right halves. From this the scaling of negativity can be inferred

$$\mathcal{N} \sim \left(L^{-(c/12)(n-1/n)} \right)^2 \Big|_{n \to \frac{1}{2}} - 1 \sim L^{c/4} \tag{4.34}$$

implying for the XX model $(c = 1)$ that

$$\mathcal{N} \sim L^{1/4}.$$

This scaling is accurately verified in the XX spin chain up to $N = 400$ spins (Fig. 4.4).

Crucially, the exponent of the power-law divergence $1/4$ does not match the earlier numerical result for the exponent $\beta = 0.47$ of our ansatz (4.24) in this model which, one might naively expect, should coincide for small ratios μ.

Firstly, we stress that the study in this section concerns finite-size scaling while our functional ansatz (4.24) has in mind the thermodynamic limit. Clearly, to

Fig. 4.4 Scaling of negativity of adjacent blocks (symmetrically bisected chain) in the XX spin chain model $(\gamma = 0, h = 0)$ as a function of the size of the blocks $L = N/2$ at the critical point in a double-logarithmic scale. The divergence of our data (*black bullets*) accurately matches with the conformal field theoretic prediction $\mathcal{N} \sim L^{c/4}$ for $c = 1$ shown as *red (gray) solid line*

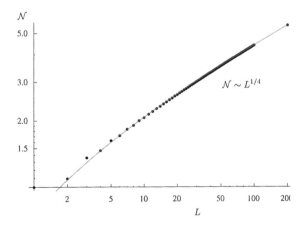

include finite size effects such as the correct scaling with L at $d = 0$ one would have to adjust the ansatz so as to comply with the study herein.

Nonetheless, the results of this section should approximately serve as an upper bound on the expected divergence for finite but small separations and large but finite block sizes, i.e. $\mu \ll 1$, since by tracing away degrees of freedom the amount of entanglement can not increase. In the case of adjacent blocks no tracing is involved which leads to this reasoning.

Hence, we conclude that our ansatz of the previous section might be too simplistic in the limit $d \to 0$ while L is large but finite.

4.6 Related Work

In Ref. [22], authors conducted independently a very similar study. These authors studied the logarithmic negativity of disjoint blocks in the continuum limit of an infinitely extended chain of coupled harmonic oscillators. They showed that logarithmic negativity displays the same qualitative feature as found for spin chains in this chapter, and that it can be accurately fitted to our ansatz (4.24) as well. Due to

$$\mathcal{L} = \ln(\mathcal{N} + 1) = \mathcal{N} - \frac{1}{2}\mathcal{N}^2 + \mathcal{O}(\mathcal{N}^3) \quad (\mathcal{N} < 1)$$

negativity and logarithmic negativity are expected to behave similarly in the relevant value range. Hence, this independent finding strongly substantiates our own results. In the case of the XX model, we fitted the logarithmic negativity to our ansatz, and we obtained $\beta \sim 1/3$, a number which has been independently confirmed by the authors in Ref. [22].

By means of bosonisation [24] the similarities of the XX spin chain model and the continuum limit of the harmonic oscillator chain become apparent: Both models are described by a bosonic field theory, yet it would be wrong to identify these fields with each other: The bosonisation approach for spins is based on the fermionisation described in Appendix A, and additionally the Kronig identity [25]

$$\sum_{k=1}^{\infty} k\hat{b}_k^\dagger \hat{b}_k = \sum_{n=1}^{\infty} n\hat{f}_n^\dagger \hat{f}_n + \frac{1}{2}\left(\hat{N}^2 + \hat{N}\right),$$

where the left hand side is a Hamiltonian of bosonic operators $\hat{b}_k$ and the right hand side is given in terms of fermionic operators $\hat{f}_n$ and the fermionic number operator $\hat{N} \equiv \sum_n \hat{f}_n^\dagger \hat{f}_n$. Close to the Fermi surface $q_k \sim \frac{\pi}{2}$ of the XX model ($N \to \infty$), one can linearise the spectrum $\epsilon_k \sim v_F |k|$ and use the Kronig relation to bosonise the theory. However, the fermion operators with respect to which the XX Hamiltonian is diagonal and which would allow the bosonisation are very non-local operators in terms of the underlying lattice. In other words, by diagonalising the spin-chain one has changed to a kind of momentum representation. On top of that, the lattice fermions are already non-local operators of the spin degrees of freedom (Sect. 2.4). Returning to a real-space coordinates within the boson picture would superficially link the harmonic oscillator chain to the bosonised spin chain. The relationship between the original spin degrees of freedom and their bosonised descendants is therefore highly nontrivial [24].

Despite this correspondence a marked difference in the value of the exponent $\alpha \sim 1$ in our study versus $\alpha \sim 2\sqrt{2}$ in their study suggests that the two models are not genuinely equivalent. The reasons for the exceptionally good match for the power-law part of the decay in both models are unknown as yet.

References

1. L. Amico, R. Fazio, A. Osterloh, V. Vedral, Entanglement in many-body systems. Rev. Mod. Phys. **80**, 517 (2009)
2. N. Goldenfeld, *Lectures on Phase Transitions and the Renormalization Group* (Perseus Books Publishing, L.L.C., Reading, 1992)
3. S. Sachdev, *Quantum Phases Transitions* (Cambridge University Press, Cambridge, 1999)
4. T.J. Osborne, M.A. Nielsen, Entanglement in a simple quantum phase transition. Phys. Rev. A **66**(3), 032110 (2002)
5. A. Osterloh, L. Amico, G. Falci, R. Fazio, Scaling of entanglement close to a quantum phase transition. Nature **416**, 608–610 (2002)
6. G. Vidal, J.I. Latorre, E. Rico, A. Kitaev, Entanglement in quantum critical phenomena. Phys. Rev. Lett. **90**(22), 227902 (2003)
7. P. Calabrese, J. Cardy. Entanglement entropy and quantum field theory. J. Stat. Mech.: Theory Exp., P06002 (2004)
8. V.E. Korepin, Universality of entropy scaling in one dimensional gapless models. Phys. Rev. Lett. **92**(9), 096402 (2004)
9. M. Henkel, *Conformal Invariance and Critical Phenomena* (Springer, Berlin, 1999)
10. H. Wichterich, J. Molina-Vilaplana, S. Bose, Scaling of entanglement between separated blocks in spin chains at criticality. Phys. Rev. A **80**, 010304(R) (2009)
11. I. Affleck, Universal term in the free energy at a critical point and the conformal anomaly. Phys. Rev. Lett. **56**(7), 746–748 (1986)
12. H.W.J. Blöte, J.L. Cardy, M.P. Nightingale, Conformal invariance, the central charge, and universal finite-size amplitudes at criticality. Phys. Rev. Lett. **56**(7), 742–745 (1986)
13. J. Eisert, M. Cramer, M.B. Plenio, Colloquium: area laws for the entanglement entropy. Rev. Mod. Phys. **82**(1), 277–306 (2010)
14. W.K. Wootters, Entanglement of formation of an arbitrary state of two qubits. Phys. Rev. Lett. **80**(10), 2245–2248 (1998)
15. T. Platini, D. Karevski, Relaxation in the xx quantum chain. J. Phys. A: Math. Theor. **40**, 1711 (2007)
16. V. Alba, L. Tagliacozzo, P. Calabrese, Entanglement entropy of two disjoint blocks in critical ising models. Phys. Rev. B **81**(6), 060411 (2010)
17. F. Iglói, I. Peschel, On reduced density matrices for disjoint subsystems. EPL (Europhys. Lett.) **89**(4), 40001 (2010)
18. I. Peschel, Calculation of reduced density matrices from correlation functions. J. Phys. A: Math. Gen. **36**(8), L205 (2003)
19. S.R. White, Phys. Rev. Lett. **69**, 2863 (1992)
20. G. Vidal, Efficient classical simulation of slightly entangled quantum computations. Phys. Rev. Lett. **91**(14), 147902 (2003)

21. J. Cardy, *Scaling and Renormalization in Statistical Physics* (Cambridge University Press, Cambridge, 1996)
22. S. Marcovitch, A. Retzker, M.B. Plenio, B. Reznik, Critical and noncritical long-range entanglement in Klein–Gordon fields. Phys. Rev. A **80**, 012325 (2009)
23. B.M. McCoy, Spin correlation functions of the $x - y$ model. Phys. Rev. **173**(2), 531–541 (1968)
24. T. Giamarchi, *Quantum Physics in One Dimension* (Clarendon Press, Oxford, 2003)
25. K. Schonhammer, V. Meden, Fermion–boson transmutation and comparison of statistical ensembles in one dimension. Am. J. Phys. **64**(9), 1168–1176 (1996)

Chapter 5
Universality of the Negativity in the Lipkin–Meshkov–Glick Model

The numerical study conducted in the previous chapter raises several questions concerning universal properties of negativity of noncomplementary regions of spins at a continuous quantum phase transition. Will this critical quantity diverge in the thermodynamic limit? Does negativity exhibit universality? If so, can its scaling features be related to known exponents of the underlying model?

Extrapolation of numerical results towards the thermodynamic may lead to rather misleading conclusions, and therefore the study of the preceding chapter can merely be considered as a hint towards universal features and scaling properties of negativity in the truly macroscopic system. The inherent complexity of the 1D models studied there has so far prevented a detailed analytical study. We argued that even if the structure of the ground state wave-function is known, it is not implied that negativity assumes a closed-form expression and, hence, its evaluation turns out to be intractable for large systems.

Building on the material of the previous chapter, our principal motivation for the work presented below is to learn about the universal properties of negativity and its scaling behaviour at quantum phase transitions in a controlled, that is, analytical fashion. This knowledge could prove helpful for hypothesising scaling relations of this quantity in the possibly less contrived critical models in 1D.

To this end, in this chapter we study the so-called Lipkin–Meshkov–Glick (LMG) model [1–3] of interacting spin-1/2 particles which becomes particularly simple in the thermodynamic limit, admitting a closed-form expression of negativity across the entire phase diagram. The simplicity of this model is bought at the expense of notions such as length, dimensionality, or region boundaries. In fact, the LMG model is defined on an infinitely coordinated graph and all coupling strengths between the constituents (spin-$\frac{1}{2}$) are of equal strength (Fig. 5.1). In order to study entanglement between noncomplementary parts of this system, a substitute for what used to be the *length* of one part will now simply be the number of spins that are contained within this region. The notion of distance between the considered parties which played a major role in the

H.C. Wichterich, *Entanglement Between Noncomplementary Parts of Many-Body Systems*, Springer Theses, DOI: 10.1007/978-3-642-19342-2_5,
© Springer-Verlag Berlin Heidelberg 2011

Fig. 5.1 Schematic of a spin system on a graph with infinite coordination. A vertex designates a single spin, and an edge denotes a uniform coupling between the vertices which it connects. The concepts of distance, dimensionality, region boundary lose their meaning in such a topology. We quantify, in this chapter, the residual entanglement between two groups of spins, e.g. those labelled 1 and 3 above, after averaging over the degrees of freedom associated to their complementary group, here 2

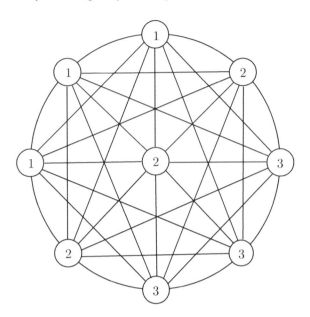

previous sections is obviously lost and will consequently be replaced by the number of spins belonging to region C (the complementary region of A and B which is not accessed by either party).

Despite its simplicity the LMG model exhibits two distinct continuous quantum phase transitions in the thermodynamic limit where the gap which separates the energy level of the ground state from that of the first excited state vanishes according to some power law upon approaching the transition. Also, a *correlation number* N_c can be defined [4] which, reminiscent of the correlation length in 1D models, exhibits an algebraic singularity at the transition.

Let us give a brief outline of this chapter. After providing the essential computational tools in Sect. 5.1 we will introduce the model under consideration and discuss its ground state properties in Sect. 5.2. We proceed by bosonising the model in Sect. 5.3, which will allow us to perform the thermodynamic limit and subsequently lead the discussion conveniently in terms of Gaussian states (Sect. 5.4). In Sect. 5.5 we report our results on logarithmic negativity between noncomplementary regions of spins in this model and discuss its properties in the vicinity of the quantum phase transition. Section 5.6 deals with the limit of complementary regions and subsequently, in Sect. 5.7, a scaling hypothesis is employed linking the finite-size scaling of logarithmic negativity with that of entropy of entanglement. Finally, we investigate the isotropic case of the LMG model in Sect. 5.8.

The work in this chapter has been done in collaboration with Dr. Julien Vidal (Université Pierre et Marie Curie, Paris, France), and with my supervisor Prof. Sougato Bose.

5.1 Mathematical Prerequisites

The aim of this section is to provide the reader with an overview of those computational tools and mathematical results of continuous variable (CV) systems and Gaussian states which are essential for the study presented in the remainder of this chapter. This section does not pretend to be an exhaustive treatment. Nowadays, this topic is a rather well established branch of quantum information and is dealt with at length in several excellent texts, notably [5–7]. Therefore, herein we adopt a utilitarian approach to most subjects and will provide details of the derivations of the relevant statements in the Appendices. We will refer to specialised literature where appropriate.

First, the notion of phase space as a symplectic vector space will be introduced. Second, the Wigner representation of the quantum state of CV systems will be recalled. Thereafter, the so-called Gaussian states will be defined in this framework and, finally, we will cover the issue of quantifying bipartite entanglement in these states in terms of (logarithmic) negativity. To that end, the essential link between partial transposition in Hilbert space and partial time-reversal [8] in phase space will be established.

5.1.1 Phase Space and Symplectic Transformations

Let $\hat{\rho}$ be a density operator on a Hilbert space $\mathcal{H}$ spanned by the direct product of Fock bases of N bosonic modes [9]

$$\mathcal{H} = \mathrm{span}(|n_1\rangle \otimes |n_2\rangle \cdots \otimes |n_N\rangle)|_{n_i=0,1,\dots}, \qquad (5.1)$$

that is $\hat{\rho}$ describes the state of $2N$ canonical degrees of freedom. We introduce the $2N$ vector $\hat{\mathbf{r}}$ of canonical operators

$$\hat{\mathbf{r}} = \begin{pmatrix} \hat{\mathbf{x}} \\ \hat{\mathbf{p}} \end{pmatrix} = [\hat{x}_1, \hat{x}_2, \dots, \hat{x}_N, \ \hat{p}_1, \hat{p}_2, \dots, \hat{p}_N]^T \qquad (5.2)$$

which is obtained from the bosonic annihilation and creation operators through

$$\begin{pmatrix} \hat{\mathbf{a}} \\ \hat{\mathbf{a}}^\dagger \end{pmatrix} = \frac{1}{\sqrt{2}} \begin{pmatrix} \mathbb{1}_N & i\mathbb{1}_N \\ \mathbb{1}_N & -i\mathbb{1}_N \end{pmatrix} \begin{pmatrix} \hat{\mathbf{x}} \\ \hat{\mathbf{p}} \end{pmatrix} \qquad (5.3)$$

in an obvious matrix notation where $\mathbb{1}_N$ denotes the $N \times N$ unit matrix. The canonical commutation relations (CCR) among these operators can be formulated in terms of the *symplectic matrix* Ω

$$\Omega_{k,l} = i[\hat{r}_k, \hat{r}_l], \quad \Omega = \begin{pmatrix} \mathbb{O}_N & -\mathbb{1}_N \\ \mathbb{1}_N & \mathbb{O}_N \end{pmatrix} \qquad (5.4)$$

with $\mathbb{O}_N$ denoting the $N \times N$ zero matrix. A homogeneous linear transformation $\mathbb{S}$ which satisfies

$$\mathbb{S}\Omega\mathbb{S}^T = \Omega \tag{5.5}$$

will be called a *symplectic transformation*. It leads to a new set of canonical operators through $\hat{\mathbf{r}} = \mathbb{S}\hat{\mathbf{r}}'$ and therefore preserves the CCR. Such transformations constitute the symplectic group [10]

Definition 5.1.1 (*Symplectic group*) The real symplectic group in $2N$ dimensions will be denoted by

$$Sp(2N, \mathbb{R}) = \{\mathbb{S} \in \mathbb{R}^{2N \times 2N} \mid \mathbb{S}\Omega\mathbb{S}^T = \Omega\}. \tag{5.6}$$

Finally, we introduce a useful theorem [11]

Theorem 5.1.2 (*Williamson*) *Let* M *be a* $2N \times 2N$ *real symmetric and positive definite matrix, then there exists a symplectic transformation* $\mathbb{S} \in Sp(2N, \mathbb{R})$ *such that*

$$\mathbb{S}M\mathbb{S}^T = \Lambda \oplus \Lambda \tag{5.7}$$

with a diagonal $N \times N$ *positive definite diagonal matrix* $\Lambda = \mathrm{diag}(\lambda_1, \lambda_2, \ldots, \lambda_N)$. *The symplectic eigenvalues* λ_n *are given by the positive square roots of the eigenvalues of* $(i\Omega M)^2$.

A simple proof, as found in Ref. [12], will be laid out in Appendix E.

5.1.2 Wigner Representation and Gaussian States

The *Wigner representation* [13] of quantum state $\hat{\rho}$ can be defined as

$$\mathcal{W}(\mathbf{x}, \mathbf{p}) = \frac{1}{(2\pi)^N} \int d^{2N}\xi\, \chi(-\xi) e^{i\xi^T \Omega \mathbf{r}}, \quad \xi = \begin{pmatrix} \xi_x \\ \xi_p \end{pmatrix} \in \mathbb{R}^{2N} \tag{5.8}$$

where $\mathbf{r} = (\mathbf{x}, \mathbf{p})^T \in \mathbb{R}^{2N}$ and $\chi(\xi)$ is the *characteristic function*

$$\chi(\xi) = \mathrm{Tr}\left[e^{i\xi^T \Omega \hat{\mathbf{r}}}\hat{\rho}\right] = \mathrm{Tr}\left[e^{i(\xi_p^T \hat{\mathbf{x}} - \xi_x^T \hat{\mathbf{p}})}\hat{\rho}\right]. \tag{5.9}$$

Within this framework, we introduce an important class of many-body states.

Definition 5.1.3 (*Gaussian N-mode state*) A quantum state $\hat{\rho}$ with characteristic function

$$\chi(\xi) = \exp\left(-\frac{1}{4}(\Omega\xi)^T \Gamma \Omega \xi - i\mathbf{d}^T \Omega \xi\right) \tag{5.10}$$

is called a Gaussian N-mode state. Here, $\mathbf{d}$ and Γ are displacement vector and covariance matrix collecting first and second moments respectively

$$\mathbf{d} \in \mathbb{R}^{2N} \qquad d_i = \langle \hat{r}_i \rangle \tag{5.11}$$

$$\Gamma \in \mathbb{R}^{2N \times 2N} \qquad \Gamma_{ij} = \langle \hat{r}_i \hat{r}_j + \hat{r}_j \hat{r}_i \rangle - 2 d_i d_j \tag{5.12}$$

where $\langle \cdots \rangle = \mathrm{Tr}(\hat{\rho} \ldots)$

We will make use of the fact that the Gaussian character of $\hat{\rho}$ is preserved under the symplectic transformation $\hat{\mathbf{r}} = \mathbb{S}\hat{\mathbf{r}}'$

$$\chi(\xi) = \mathrm{Tr}\left[\hat{\rho} e^{i\xi^T \Omega \hat{\mathbf{r}}}\right] = \mathrm{Tr}\left[\hat{\rho} e^{i\xi^T \Omega \mathbb{S} \hat{\mathbf{r}}'}\right] = \mathrm{Tr}\left[\hat{\rho} e^{i(\Omega \mathbb{S}^T \Omega^T \xi)^T \Omega \hat{\mathbf{r}}'}\right] \tag{5.13}$$

and with $\mathbb{S}^{-1} = \Omega \mathbb{S}^T \Omega^T$ it follows that

$$\chi(\xi) = \chi'(\mathbb{S}^{-1} \xi). \tag{5.14}$$

It can be straight-forwardly shown that the covariance matrices in both representations are related through

$$\Gamma = \mathbb{S} \Gamma' \mathbb{S}^T \tag{5.15}$$

Finally, it is easy to see that the reduced density matrices, $\hat{\rho}_S = \mathrm{Tr}_E \hat{\rho}$, which arise from $\hat{\rho}$ through partial trace Tr_E over environmental degrees of freedom are Gaussian states if $\hat{\rho}$ is a Gaussian state. The covariance matrix Γ_S of the reduced state $\hat{\rho}_S$ is obtained by erasing those rows and columns of Γ associated to the environment.

5.1.3 Partial Transposition and Bipartite Entanglement of Gaussian States

The set of all canonical degrees of freedom $\{\hat{x}_i, \hat{p}_i\}, i = 1, \ldots, N$ may be divided into a subset $\{\hat{x}_i, \hat{p}_i\} i = 1, \ldots, N_A$ associated to Alice (A) and its complement which we assume is belonging to Bob (B). The entanglement that is shared between A and B, assuming their global state $\hat{\rho}_{AB}$ is a Gaussian state *which need not be pure*, will be quantified in terms of logarithmic negativity $\mathcal{L} = \ln \|\hat{\rho}_{AB}^{T_A}\|$. The following alternative definition of the Wigner function (see Appendix F for the connection to definition (5.8) and further details)

$$W(\mathbf{x}, \mathbf{p}) = \frac{1}{(2\pi)^N} \int d^N \xi_x \, \langle \mathbf{x} - \xi_x/2 | \hat{\rho}_{AB} | \mathbf{x} + \xi_x/2 \rangle e^{-i\xi_x^T \mathbf{p}}, \quad \xi_x \in \mathbb{R}^N$$

elucidates, by inspection, that under partial transposition T_A the Wigner function is subjected to a reversal of the momenta belonging to Alice $\mathbf{p}_A \to -\mathbf{p}_A$ (a.k.a.

partial time-reversal) [8]. Now, let Γ be the covariance matrix of the Gaussian state $\hat{\rho}_{AB}$. The correspondence that was set forth above implies that $\hat{\rho}_{AB}^{T_A}$ has, again, a Gaussian characteristic function with covariance matrix Γ^{T_A} arising from Γ by changing the sign in those elements $\Gamma_{i,j}$ which involve an unpaired momentum variable of Alice's (*caution: no actual transposition is involved on the level of covariance matrices despite the notation* Γ^{T_A}). Logarithmic negativity assumes (Appendix I)

$$\mathcal{L} = \ln \|\hat{\rho}_{AB}^{T_A}\| = -\sum_{\lambda_m < 1} \ln \lambda_m. \qquad (5.16)$$

in terms of the symplectic eigenvalues λ_m of Γ^{T_A} (see Appendix I for a derivation).

5.2 The Model and the Hamiltonian

The LMG model describes a collection of mutually interacting spin-$\frac{1}{2}$ and is well-established in the study of magnetic phase transitions in nuclei [1–3]. In terms of spin-$\frac{1}{2}$ operators its Hamiltonian reads as

$$\hat{H} = -\frac{1}{2N} \sum_{i<j} \left(\hat{\sigma}_i^x \hat{\sigma}_j^x + \gamma \hat{\sigma}_i^y \hat{\sigma}_j^y \right) - \frac{h}{2} \sum_i \hat{\sigma}_i^z - \frac{1}{4}(1+\gamma) \qquad (5.17)$$

highlighting the similarities to the one-dimensional XY model from the previous sections, except for the infinite coordination (all spins are coupled mutually, see Fig. 5.1 for an illustration) and a factor $1/N$ which ensures that the ground state energy per spin remains finite in the thermodynamic limit [14].

In the bosonisation approach to be laid out below, and in the thermodynamic limit, one can distinguish two distinct magnetic phases. For $h > 1$ (symmetric phase) the ground state is unique and a gap $\Delta = \sqrt{(h-1)(h-\gamma)}$ separates its energy from that of the first excited state. Furthermore the magnetisation is aligned with the external magnetic field. In the so called broken phase $h < 1$ a two-fold degeneracy is predicted to develop as $\sim e^{-N}$ [4], and is therefore manifest in the thermodynamic limit. The two degenerate ground states are eigenstates of the parity (or spin-flip) operator $\hat{Z} = \prod_{k=1}^{N} \hat{\sigma}_k^z$, so for finite N this spin-flip symmetry is broken, and the ground state has a definite parity ± 1. In the present study, which mainly concerns the thermodynamic limit, we will focus on one of these degenerate ground states which—by virtue of continuity—originates from its unique, finite N analogue [14].

Importantly, continuous phase transitions occur at $h = h_c = 1$ and also in the isotropic limit $\gamma = 1$ where in both cases the gap closes according to a power law in N [14]. Distinct sets of critical exponents which are associated to these transitions allow a comparative study of universal behaviour of physical quantities.

The present study exploits this property and aims at the investigation of universality of logarithmic negativity.

A more convenient form of (5.17) for the subsequent discussion is achieved in terms of collective spin operators

$$\hat{H} = -\frac{1}{4N}\sum_{ij}\left(\hat{\sigma}_i^x\hat{\sigma}_j^x + \gamma\hat{\sigma}_i^y\hat{\sigma}_j^y\right) - h\hat{S}_z \tag{5.18}$$

$$\hat{H} = -\frac{1}{N}\left(\hat{S}_x^2 + \gamma\hat{S}_y^2\right) - h\hat{S}_z \tag{5.19}$$

$$\hat{H} = -\frac{1}{2N}(1+\gamma)\left(\hat{S}^2 - \hat{S}_z^2\right) - \frac{1}{4N}(1-\gamma)\left(\hat{S}_+^2 + \hat{S}_-^2\right) - h\hat{S}_z. \tag{5.20}$$

Collective spin operators are defined in the usual way

$$\hat{S}_\alpha \equiv \frac{1}{2}\sum_i \hat{\sigma}_i^\alpha \quad (\alpha = x, y, z) \tag{5.21}$$

$$\hat{S}^2 \equiv \hat{S}_x^2 + \hat{S}_y^2 + \hat{S}_z^2 \tag{5.22}$$

$$\hat{S}_\pm \equiv \hat{S}_x \pm i\hat{S}_y. \tag{5.23}$$

From (5.20) one recognises that for every value of γ

$$[\hat{S}^2, \hat{H}] = [\hat{S}^2, \hat{S}_z] = [\hat{S}^2, \hat{S}_\pm] = 0.$$

Therefore, the ground state is a superposition of the permutation symmetric Dicke states which are defined as (assume N is even for simplicity)

$$\hat{S}^2|S, M\rangle = S(S+1)|S, M\rangle \quad S = 0, 1, \ldots, N/2 \tag{5.24}$$

$$\hat{S}_z|S, M\rangle = M|S, M\rangle \quad M = -S, -S+1, \ldots, S \tag{5.25}$$

and consequently the ground state belongs to the maximum spin sector $S = N/2$ which minimises the energy. This also renders numerical studies in this model very convenient, since the representation of Hamiltonian (5.20) can be restricted to this relevant $N + 1$ dimensional spin sector.

5.3 Mapping to a Three Mode Boson Problem

The collection of spins can be partitioned arbitrarily into groups of N_1, N_2 and N_3 spins respectively, so that $N_1 + N_2 + N_3 = N$. This will allow us to study entanglement between two of the parties while the degrees of freedom belonging to the third party will be traced over. One decomposes collective spin operators accordingly into $S_\alpha = S_\alpha^{(1)} + S_\alpha^{(2)} + S_\alpha^{(3)}$ and proceeds by casting the Hamiltonian

(5.20) into this partitioned representation. Concurrently, we bosonise the expressions making use of

$$\hat{S}_z^{(k)} = S^{(k)} - \hat{a}_k^\dagger \hat{a}_k \tag{5.26}$$

$$\hat{S}_-^{(k)} = (\hat{S}_+^{(k)})^\dagger = \hat{a}_k^\dagger (2S^{(k)} - \hat{a}_k^\dagger \hat{a}_k)^{1/2}, \quad k = 1, 2, 3 \tag{5.27}$$

where operators $\hat{a}_k$ obey the bosonic commutation relations

$$[\hat{a}_k^\dagger, \hat{a}_l] = \delta_{kl}, \quad [\hat{a}_k^\dagger, \hat{a}_l^\dagger] = [\hat{a}_k, \hat{a}_l] = 0.$$

The mapping in equations (5.26–5.27) is known as the Holstein–Primakoff representation [15]. This procedure casts the spin problem into the more convenient language of three bosonic modes, each of which represents a group of spins. Since in the ground state one has $S^{(k)} = N_k/2$, we neglect terms of order higher than $\mathcal{O}(1/N)^0$ which is justified for $N \to \infty$. In taking this limit, we assume that the ratios $\tau_k = N_k/N$, $k = 1, 2, 3$ are fixed. Additionally, in the broken phase $h < 1$ it is necessary to rotate the spin representation such that the z component coincides with the magnetisation axis [14]. The detailed calculation is carried out in Appendix K. Omitting scalar valued terms, which will be irrelevant in the following, the Hamiltonian assumes

$$\hat{H} \sim \sum_{k,l=1}^3 \hat{a}_k^\dagger \mathbb{A}_{k,l} \hat{a}_l + \frac{1}{2}(\hat{a}_k^\dagger \mathbb{B}_{k,l} \hat{a}_l^\dagger + \text{H.c.}) + \mathcal{O}(N^{-1}). \tag{5.28}$$

Recall that the bosonic operators $\hat{a}_k^\dagger$ and $\hat{a}_k$ are labelled according to the group of spins from which they originate. We have introduced the adjacency matrices

$$\mathbb{A} = r \begin{pmatrix} 1 & 0 & 0 \\ 0 & 1 & 0 \\ 0 & 0 & 1 \end{pmatrix}, \quad r = \begin{cases} \frac{2h-\gamma-1}{2} & h \geq 1, \\ \frac{2-\gamma-h^2}{2} & 0 \leq h < 1, \end{cases} \tag{5.29}$$

$$\mathbb{B} = s \begin{pmatrix} \tau_1 & \sqrt{\tau_1\tau_2} & \sqrt{\tau_1\tau_3} \\ \sqrt{\tau_1\tau_2} & \tau_2 & \sqrt{\tau_2\tau_3} \\ \sqrt{\tau_1\tau_3} & \sqrt{\tau_2\tau_3} & \tau_3 \end{pmatrix}, \quad s = \begin{cases} \frac{\gamma-1}{2} & h \geq 1, \\ \frac{\gamma-h^2}{2} & 0 \leq h < 1. \end{cases} \tag{5.30}$$

Let us finally note that $r > 0$ and $r > s$ in both broken and symmetric phase. Equation (5.28) is a quadratic form, that can be diagonalised by a canonical transformation to new bosonic operators $\hat{\eta}_m$ such that

$$H \sim \sum_m \epsilon_m \left(\hat{\eta}_m^\dagger \hat{\eta}_m + \frac{1}{2}\right) \tag{5.31}$$

where ϵ_m denote the single particle energies. Hence, if all $\epsilon_m > 0$, the ground state is non-degenerate and coincides with the bosonic vacuum $\eta_m|0\rangle = 0 \forall m$.

5.4 Derivation of the Covariance Matrix

As shown in Appendix G, the bosonic vacuum $|0\rangle$ is a Gaussian state with covariance matrix $\Gamma' = \mathbb{1}$ and vanishing first moments. In order to obtain the covariance matrix Γ of the ground state of (5.28) in the original coordinates, one invokes formula (5.15) where $\mathbb{S}$ takes the role of the symplectic transformation which gives rise to the diagonal form of (5.28). This rather technical step will be revised in Appendix H. One finds [16, 17]

$$\Gamma = \Gamma_x \oplus \Gamma_p \tag{5.32}$$

$$\Gamma_x = \Gamma_p^{-1} = V_x^{-1/2}(V_x^{1/2}V_pV_x^{1/2})^{1/2}V_x^{-1/2} \tag{5.33}$$

$$V_x = \mathbb{A} + \mathbb{B} \geq 0 \tag{5.34}$$

$$V_p = \mathbb{A} - \mathbb{B} \geq 0. \tag{5.35}$$

The direct sum in (5.32) leads to a block-diagonal covariance matrix, therefore second moments of the form $\langle x_i p_j\rangle + \langle p_j x_i\rangle$ which mix coordinates and momenta vanish identically. Since in our special case one has that, trivially, $[\mathbb{A}, \mathbb{B}] = 0$ it follows that $[V_x, V_p] = 0$ and consequently the covariance matrix can be further simplified, yielding

$$\Gamma_x = \Gamma_p^{-1} = V_p^{1/2}V_x^{-1/2}. \tag{5.36}$$

Combining (5.29) and (5.30) with (5.36) and noting that $\mathbb{B}^n = s^{n-1}\mathbb{B}$ we find

$$
\begin{aligned}
\Gamma_x &= \left[\mathbb{1} - \frac{\mathbb{B}}{r}\right]^{1/2}\left[\mathbb{1} + \frac{\mathbb{B}}{r}\right]^{-1/2} \\
&= \left[\sum_{k=0}^{\infty}\binom{1/2}{k}\left(-\frac{\mathbb{B}}{r}\right)^k\right]\left[\sum_{l=0}^{\infty}\binom{-1/2}{l}\left(\frac{\mathbb{B}}{r}\right)^l\right] \\
&= \left[\mathbb{1} + \frac{\mathbb{B}}{s}\sum_{k=1}^{\infty}\binom{1/2}{k}\left(-\frac{s}{r}\right)^k\right]\left[\mathbb{1} + \frac{\mathbb{B}}{s}\sum_{l=1}^{\infty}\binom{-1/2}{l}\left(\frac{s}{r}\right)^l\right] \\
&= \left[\mathbb{1} + \frac{\mathbb{B}}{s}\left(\left(1 - \frac{s}{r}\right)^{1/2} - 1\right)\right]\left[\mathbb{1} + \frac{\mathbb{B}}{s}\left(\left(1 + \frac{s}{r}\right)^{-1/2} - 1\right)\right] \\
&= \mathbb{1} + \left(\sqrt{\frac{r-s}{r+s}} - 1\right)\frac{\mathbb{B}}{s} = \mathbb{1} + (\alpha^{-1} - 1)\frac{\mathbb{B}}{s}
\end{aligned}
\tag{5.37}
$$

where in the second line we invoked the binomial theorem and we introduced the short $\alpha = \sqrt{(r+s)/(r-s)}$. An analogous computation leads to

$$\Gamma_p = \mathbb{1} + (\alpha - 1)\frac{\mathbb{B}}{s}. \tag{5.38}$$

which concludes the derivation of the covariance matrix.

5.5 Logarithmic Negativity in a Tripartition

Let us return to our objective of computing entanglement of noncomplementary groups of spins in this model. As we saw in Sect. 5.1 the logarithmic negativity between two groups, 1 and 3, say, of spins which persists after averaging over the degrees of freedom of the remainder, here 2, can now be computed from the 4×4 submatrix $\widetilde{\Gamma}$ of Γ which is obtained by keeping only rows and columns of Γ that are belonging to the considered modes (1 and 3). In the present representation the submatrix assumes $\widetilde{\Gamma} = \widetilde{\Gamma}_x \oplus \widetilde{\Gamma}_p$ where

$$\widetilde{\Gamma}_x = \mathbb{1} + (\alpha^{-1} - 1)\begin{pmatrix} \tau_1 & \sqrt{\tau_1 \tau_3} \\ \sqrt{\tau_1 \tau_3} & \tau_3 \end{pmatrix} \tag{5.39}$$

$$\widetilde{\Gamma}_p = \mathbb{1} + (\alpha - 1)\begin{pmatrix} \tau_1 & \sqrt{\tau_1 \tau_3} \\ \sqrt{\tau_1 \tau_3} & \tau_3 \end{pmatrix} \tag{5.40}$$

We proceed with determining the Simon invariants (see Sect. 5.1) of $\widetilde{\Gamma}$ in terms of which the symplectic spectra of $\widetilde{\Gamma}$ and $\widetilde{\Gamma}^{T_1}$ can be computed conveniently. To this end, we reshuffle the canonical variables so as to group degrees of freedom belonging to the individual parties $r = (x_1, x_3, p_1, p_3)^T \rightarrow r' = (x_1, p_1, x_3, p_3)^T$ leading to

$$\widetilde{\Gamma} = \begin{pmatrix} P & R \\ R & Q \end{pmatrix}$$

$$P = \mathbb{1} + \tau_1 \begin{pmatrix} \alpha^{-1} - 1 & 0 \\ 0 & \alpha - 1 \end{pmatrix}, \quad \det P = 1 + \tau_1(1 - \tau_1)(\alpha + \alpha^{-1} - 2)$$

$$Q = \mathbb{1} + \tau_3 \begin{pmatrix} \alpha^{-1} - 1 & 0 \\ 0 & \alpha - 1 \end{pmatrix}, \quad \det Q = 1 + \tau_3(1 - \tau_3)(\alpha + \alpha^{-1} - 2)$$

$$R = \sqrt{\tau_1 \tau_3} \begin{pmatrix} \alpha^{-1} - 1 & 0 \\ 0 & \alpha - 1 \end{pmatrix}, \quad \det R = -\tau_1 \tau_3(\alpha + \alpha^{-1} - 2).$$

Partial time-reversal (Appendix F) amounts to $\det R \rightarrow \det R' = -\det R$, and the characteristic polynomial for the matrix[1] $(i\Omega \widetilde{\Gamma}^{T_1})^2$ can now be written in terms of $\det P$, $\det Q$, $\det R$, and $\det \widetilde{\Gamma}$ [8, 18, 19]

$$\lambda^4 + (\det P + \det Q - 2 \det R)\lambda^2 + \det \widetilde{\Gamma} = \tag{5.41}$$

$$\lambda^4 + 2(1 + g)\lambda^2 + (1 + 2g') = 0 \tag{5.42}$$

where

---

[1] Recall that the eigenspectrum of this matrix leads to the symplectic spectrum of Γ^{T_1} by virtue of Theorem 5.1.2

$$g = \frac{1}{2}\left(\tau_1 + \tau_3 - (\tau_1 - \tau_3)^2\right)(\alpha + \alpha^{-1} - 2) \tag{5.43}$$

$$g' = \frac{1}{2}\left(\tau_1 + \tau_3 - (\tau_1 + \tau_3)^2\right)(\alpha + \alpha^{-1} - 2) \tag{5.44}$$

leading to the symplectic eigenvalues λ_+ and λ_-

$$\lambda_\pm = \sqrt{1 + g \pm \sqrt{(1+g)^2 - (1+2g')}} \tag{5.45}$$

$$\lambda_\pm = \sqrt{1 + g \pm \sqrt{g^2 + 4\tau_1\tau_3(\alpha + \alpha^{-1} - 2)}}. \tag{5.46}$$

Clearly, we have that $\lambda_+ \geq 1$ and $\lambda_- \leq 1$ so that by virtue of (5.16) logarithmic negativity is simply given by

$$\mathcal{L}(\tau_1, \tau_3, \gamma, h) = -\ln\lambda_- \tag{5.47}$$

which is the main result of this chapter (Fig. 5.2). Let us finally discuss the limiting value of $\mathcal{L}$ when $h \to 1$, that is the system is at criticality. Equivalently we may study $\alpha \to 0$, thereby saving ourselves studying the limits $h \to 1^+$ and $h \to 1^-$ separately. Departing from (5.46) and introducing the shorthand notations $x = (\alpha + \alpha^{-1} - 2)$ and $y = (\tau_1 + \tau_3 - (\tau_1 - \tau_3)^2)/2$ we find

$$\lim_{\alpha \to 0} \lambda_- = \lim_{x \to \infty} \sqrt{1 + xy - \sqrt{x^2y^2 + 4\tau_1\tau_3 x}} \tag{5.48}$$

$$\lim_{\alpha \to 0} \lambda_- = \lim_{x \to \infty} \sqrt{1 + xy\left(1 - \sqrt{1 + \frac{4\tau_1\tau_3}{xy^2}}\right)} \tag{5.49}$$

$$\lim_{\alpha \to 0} \lambda_- = \lim_{x \to \infty} \sqrt{1 + xy\left(1 - 1 - \frac{2\tau_1\tau_3}{xy^2}\right)} \tag{5.50}$$

$$\lim_{\alpha \to 0} \lambda_- = \sqrt{\frac{\tau_1 + \tau_3 - (\tau_1 + \tau_3)^2}{\tau_1 + \tau_3 - (\tau_1 - \tau_3)^2}} \tag{5.51}$$

and therefore (5.47) assumes[2]

$$\mathcal{L}(\tau_1, \tau_3, \gamma, h = 1) = -\frac{1}{2}\ln\left(\frac{\tau_1 + \tau_3 - (\tau_1 + \tau_3)^2}{\tau_1 + \tau_3 - (\tau_1 - \tau_3)^2}\right). \tag{5.52}$$

---

[2] we stress that there is a typo in (5.12) of [20] and should read as (5.52) of this treatise.

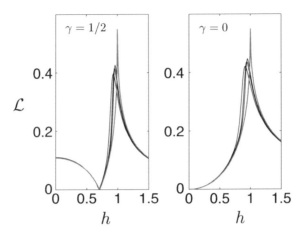

Fig. 5.2 Logarithmic negativity $\mathcal{L}$ for an equal tripartition $\tau_1 = \tau_2 = \tau_3 = 1/3$ as a function of the magnetic field h and for two different values of the anisotropy parameter γ. In the thermodynamic limit, corresponding to the *red (gray) lines*, at the critical point, $\mathcal{L}$ is universal (independent of γ). *Black lines* correspond to numerical data for $N = 90, 150, 210$, which match the analytical prediction $N = \infty$ increasingly well. Note also that $\mathcal{L}$ vanishes for $h = \sqrt{\gamma}$ where the ground state is separable [14]. Figure reprinted from [20]

We see that $\mathcal{L}$ assumes a finite value in the thermodynamic limit and for any tripartition $0 < \tau_1, \tau_2, \tau_3 < 1$ such that $\tau_1 + \tau_2 + \tau_3 = 1$, including the critical point $h = 1$ where it is manifestly independent of the anisotropy γ.

5.6 Logarithmic Negativity in a Bipartition

The limit of $\tau_2 \to 0$ corresponds to the bipartite case, a situation where negativity between parties 1 and 3 is singular at $h = 1, 0 \le \gamma < 1$. In order to study this divergence, one should impose a bipartition from the start since we required all ratios $\tau_k, k = 1, 2, 3$ to be fixed and nonzero when taking the thermodynamic limit. Then, studying $\mathcal{L}$ in the vicinity of $\tau_2 \sim 0$ is certainly not allowed and may lead to wrong implications (recall that we expanded H to zeroth order of $1/N_2$). However, nothing prevents us from studying the negativity between party 1 and $2 \cup 3$ which is what we do now. So the common state of the considered parties is no longer mixed but pure, leading to a drastic simplification. In fact, as shown in Sect. 1.1 (compare (1.28)), the logarithmic negativity can be obtained, in this particular case, in terms of the eigenvalues $w_n, n = 0, 1, 2 \ldots$ of the reduced density matrix $\hat{\rho}_1$ of subsystem 1 (obtained from tracing away 2 and 3)

$$\mathcal{L} = 2 \ln \sum_n \sqrt{w_n} \tag{5.53}$$

The covariance matrix of $\hat{\rho}_1$ becomes

$$\tilde{\Gamma} = \begin{pmatrix} 1 + (\alpha^{-1} - 1)\tau_1 & 0 \\ 0 & 1 + (\alpha - 1)\tau_1 \end{pmatrix}. \tag{5.54}$$

Even though $\tilde{\Gamma}$ is diagonal, the diagonal entries are *not* the symplectic eigen-values. Williams theorem implies that we have only one symplectic eigenvalue which is obtained from the spectrum of $(i\Omega\tilde{\Gamma})^2$ and reads as

$$\lambda = \sqrt{(1 + (\alpha^{-1} - 1)\tau_1)(1 + (\alpha - 1)\tau_1)}. \tag{5.55}$$

The eigenvalues of $\hat{\rho}_1$ are related to λ as (compare J.12 of Appendix J)

$$w_n = \frac{2}{\lambda + 1}\left(\frac{\lambda - 1}{\lambda + 1}\right)^n, \quad n = 0, 1, 2, \dots \tag{5.56}$$

Logarithmic negativity becomes, using (5.53)

$$\mathcal{L}(\tau_1, 1 - \tau_1, \gamma, h) = 2\ln\sqrt{\frac{2}{\lambda + 1}}\sum_n\left(\sqrt{\frac{\lambda - 1}{\lambda + 1}}\right)^n \tag{5.57}$$

$$\mathcal{L}(\tau_1, 1 - \tau_1, \gamma, h) = \ln\frac{2}{\left(\sqrt{\lambda + 1} - \sqrt{\lambda - 1}\right)^2}. \tag{5.58}$$

For fixed values of $\tau_1 = 1/3$, $\gamma = 1/2$ we plot $\mathcal{L}$ as a function of h in Fig. 5.3 and our result will be studied near criticality in the following chapter.

5.7 Finite-Size Scaling

In principle, phase transitions occur only in the thermodynamic limit $N \to \infty$. Phenomenological finite-size scaling (FSS) is based upon the insight that while certain physical quantities ϕ diverge according to some power law in the ther-modynamic limit and upon approaching the transition $h \to h_c$

$$\phi \sim |h - h_c|^{\xi_\phi} \quad (N = \infty). \tag{5.59}$$

Here, h denotes a parameter that drives the transition such that the system is critical for $h = h_c$ and $N = \infty$. In finite systems no such thermodynamic singu-larity can occur and the quantities ϕ are well-behaved and finite: The limiting critical behaviour is inhibited. Principal objects of FSS are either, given the properties in the thermodynamic limit, to find the dependence of ϕ on N at the point $h = h_c$ or, conversely, given the scaling with N (e.g., from numerics) in the vicinity of an anticipated $h = h_c$, to infer about model properties at $N = \infty$ which

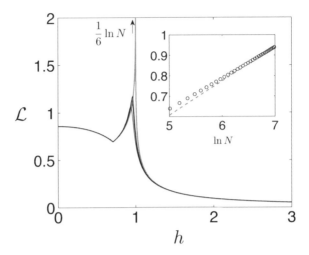

Fig. 5.3 Logarithmic negativity $\mathcal{L}$ as a function of the magnetic field h for a bipartition $\tau_1 = 1/3$, $\tau_3 = 2/3$, and $\gamma = 1/2$, shown as *red (gray) line*. In the broken phase $h < 1$ numerical data display an artificial shift of $\ln 2$ in $\mathcal{L}$ which is a signature of the parity-broken ground state [21]. We account for this shift by plotting $\mathcal{L} + \ln 2$ in this regime (at $h = \sqrt{\gamma}$ the ground state is separable [21, 22]). *Black lines* corresponds to exact diagonalisation data for $N = 180, 270, 360$. Inset: Scaling of $\mathcal{L}$ with system size $N = 120, 150, \ldots, 1080$ from exact diagonalisation (*black circles*) approaching a linear dependence on $\ln N$ with slope $1/6$ (*dashed line*) for large N. Figure reprinted from [20]

would otherwise be inaccessible. To this end, and in order to comply with the insight stated above, one hypothesises the existence of a regular function G_ϕ called *scaling function* such that in the vicinity of the transition ϕ obeys a so-called *scaling form*

$$\phi \sim |h - h_c|^{\xi_\phi} G_\phi(N|h - h_c|^{\nu^*}). \qquad (5.60)$$

Here, the exponents ξ_ϕ and n_ϕ are specific to the quantity ϕ under consideration while ν^* is model specific [4]. The properties of the scaling function are assumed to be such that $G_\phi(x) \rightarrow$ const for $x \rightarrow \infty$ granting that the divergent behaviour (5.59) is recovered when $N \rightarrow \infty$. When $x \rightarrow 0$, corresponding to finite N and $|h - h_c| \rightarrow 0$, one supposes that $G_\phi(x) \sim x^{-\xi_\phi/\nu^*}$. In this way, G_ϕ leads to an identical cancellation of the divergent term and ϕ remains finite, implying a non-trivial scaling of ϕ with N as

$$\phi \sim |h - h_c|^{\xi_\phi} (N|h - h_c|^{\nu^*})^{-\xi_\phi/\nu^*} = N^{-\xi_\phi/\nu^*} \quad (h = h_c). \qquad (5.61)$$

In order to apply these ideas to logarithmic negativity of a fixed bipartition of the LMG model, at first we note that one identifies $\phi = e^{\mathcal{L}}$ as a quantity which diverges as a power law at the transition. Further, the scaling form (5.60) with $\nu^* = 3/2$ was verified with high accuracy through explicit expansion in $1/N$ of many different quantities in this model. We proceed by expanding[3] $e^{\mathcal{L}}$ from (5.57) around $h = 1^{+}$ revealing that at the leading order

$$e^{\mathcal{L}} \sim \frac{2(1-\gamma)^{1/4}(\tau_1(1-\tau_1))^{1/2}}{|h-1|^{1/4}} + \mathcal{O}(|h-1|^{1/4})$$

and hence $\xi_\phi = -1/4$. Therefore, we conclude that $\phi \sim N^{1/6}$ or

$$\mathcal{L}(\tau_1, 1-\tau_1, \gamma, h = 1) \sim \frac{1}{6}\ln N \tag{5.62}$$

which is the result of this section. The same scaling behaviour was found for other measures in this model, including entropy of entanglement $\mathcal{E}$, geometric entanglement $\mathcal{G}$, and single-copy entanglement $\mathcal{S}$ which leads us to conjecture that a similar equivalence holds in other models, including one-dimensional spin chains for which the equivalence among $\mathcal{E}$, $\mathcal{G}$, and $\mathcal{S}$ has already been established [23].

It would be desirable to test the scaling hypothesis, for example, by computing the leading terms of an $1/N$ expansion of ϕ, similar to what was done by Barthel et al. in [22]. In this way one could verify the particular form of the scaling variable that was hypothesised above. Given the considerable complication that a $1/N$ expansion would entail (keeping higher order terms of the Holstein Primakoff boson mapping in Sect. 5.3), we choose to verify the predicted size scaling of $\mathcal{L}$ numerically, instead. This is done by way of exact diagonalisation in systems of up to $N \sim 1000$ spins (see Fig. 5.3).

As a final remark let us note that the origin of the logarithmic divergence above is most probably different from the ubiquitous $\ln l$ scaling of $\mathcal{E}$ [24] in one-dimensional models at criticality (where l is the length of the subsystem). In the present study, the fact that one will observe at most logarithmic divergence with the subsystem size can be traced back to the symmetry properties of the ground state of the LMG model. As argued in Sect. 5.2 the ground state of the LMG model is a superposition of Dicke states

$$|\phi_0\rangle = \sum_{M=-N/2}^{N/2} p_M |N/2, M\rangle \quad (N \text{ even}).$$

The Schmidt decomposition of Dicke states, that will be introduced in the following section (5.70), implies that the Schmidt rank χ of $|\phi_0\rangle$ for an equal bipartition is at most $N/2 + 1$. This leads to an upper bound on $\mathcal{L}$ (which is achieved when all Schmidt numbers $\sqrt{w_n}$ are equal and the Schmidt rank assumes its maximum value)

---

[3] The leading term of this expansion is found with the aid of computer algebra software Mathematica.

$$\mathcal{L} = 2 \ln \left(\sum_{n=1}^{\chi} \sqrt{w_n} \right) \tag{5.63}$$

$$\mathcal{L} \le 2 \ln \left(\sum_{k=1}^{N/2+1} \frac{1}{\sqrt{N/2+1}} \right) \tag{5.64}$$

$$\mathcal{L} = \ln(N/2 + 1) \sim \ln N \quad (N \to \infty) \tag{5.65}$$

One has additionally the lower bound $\mathcal{E} \le \mathcal{L}$ for pure states (1.29). In the LMG model it was found that $\mathcal{E} \sim 1/6 \ln N$ [22] and consequently

$$1/6 \ln N \le \mathcal{L} \le \ln N.$$

5.8 The Isotropic Case

The final step in giving a rather complete picture of negativity in the LMG model will be to look at the isotropic case $\gamma = 1$, which has been disregarded so far. The model is again critical on the entire line $\gamma = 1, 0 \le h \le 1$ but belongs to a different universality class compared to the transition studied in the foregoing sections. For $\gamma = 1$ the Hamiltonian (5.20) commutes with $\mathbf{S}^2$ as well as $\hat{S}_z$ which implies that the ground states are the Dicke states $|S, M\rangle$ which were defined in Equations (5.24) and (5.25). The quantum number M changes as a function of h, and in the limit $h = 1^-$ the ground state is $|N/2, N/2 - 1\rangle$, that is, a permutation symmetric superposition of a single spin deviation in an otherwise fully polarised background of spins. So we can visualise the presence or absence of that single spin with $|\uparrow\rangle, |\downarrow\rangle$ This state reads in an *equal* tripartition $\tau_1 = \tau_2 = \tau_3 = 1/3$

$$|N/2, N/2 - 1\rangle = \frac{1}{\sqrt{3}}(|\uparrow\rangle|\downarrow\rangle|\downarrow\rangle + |\downarrow\rangle|\uparrow\rangle|\downarrow\rangle + |\downarrow\rangle|\downarrow\rangle|\uparrow\rangle) \tag{5.66}$$

for which the negativity may be worked out straight forwardly. The reduced density operator of two of the groups, and its partial transpose, assume

$$\hat{\rho}_{\text{red}} = \frac{1}{\sqrt{3}} \begin{pmatrix} 1 & 0 & 0 & 0 \\ 0 & 1 & 1 & 0 \\ 0 & 1 & 1 & 0 \\ 0 & 0 & 0 & 0 \end{pmatrix} \tag{5.67}$$

$$\hat{\rho}_{\text{red}}^{T_2} = \frac{1}{\sqrt{3}} \begin{pmatrix} 1 & 0 & 0 & 1 \\ 0 & 1 & 0 & 0 \\ 0 & 0 & 1 & 0 \\ 1 & 0 & 0 & 0 \end{pmatrix} \tag{5.68}$$

The negative eigenvalue of $\hat{\rho}_{\text{red}}$ reads $\lambda_{-} = \frac{1}{2\sqrt{3}}(1 - \sqrt{5})$ and therefore logarithmic negativity in this setting is

$$\mathcal{L} = \ln\left(\frac{1}{\sqrt{3}}(\sqrt{5} - 1) + 1\right) \tag{5.69}$$

Let us now look at a *general* tripartition of the state $|N/2, N/2 - 1\rangle$. The Dicke state $|N/2, M\rangle$ of N spin-$\frac{1}{2}$ particles can be written in a bipartite fashion (dividing the system into N_1 and N_2 spins with $N_1 + N_2 = N$ assuming N_1 and N_2 are even integers)

$$\left|\frac{N}{2}, M\right\rangle = \left|\frac{N}{2}, \frac{N}{2} - n\right\rangle = \sum_k \sqrt{\frac{\binom{N_1}{k}\binom{N_2}{n-k}}{\binom{N}{n}}} \left|\frac{N_1}{2}, \frac{N}{2} - k\right\rangle \left|\frac{N_2}{2}, \frac{N}{2} - n + k\right\rangle \tag{5.70}$$

where the sum runs over allowed values of k commensurate with the relationship between quantum numbers S and M appearing on the right hand side ($M \in \{-S, -S+1, \ldots, S\}$). This is the Schmidt decomposition of this particular Dicke state which can be found in, e.g. [25]. Performing this decomposition twice, the ground state can be written in tripartite fashion as

$$\left|\frac{N}{2}, \frac{N}{2} - 1\right\rangle = \sum_{k=0}^{1} \sum_{l=0}^{1-k} \sqrt{\frac{\binom{\tau_1 N}{k}\binom{\tau_2 N}{l}\binom{\tau_3 N}{1-k-l}}{\binom{N}{1}}}$$
$$\times \left|\frac{\tau_1 N}{2}, \frac{N}{2} - k\right\rangle \left|\frac{\tau_2 N}{2}, \frac{N}{2} - l\right\rangle \left|\frac{\tau_3 N}{2}, \frac{N}{2} - 1 + l + k\right\rangle \tag{5.71}$$

$$\left|\frac{N}{2}, \frac{N}{2} - 1\right\rangle = \sqrt{\tau_1}|\uparrow\rangle|\downarrow\rangle|\downarrow\rangle + \sqrt{\tau_2}|\downarrow\rangle|\uparrow\rangle|\downarrow\rangle + \sqrt{\tau_3}|\downarrow\rangle|\downarrow\rangle|\uparrow\rangle \tag{5.72}$$

The reduced density operator of partitions 1 and 3 (after tracing away 2) reads

$$\hat{\rho}_{\text{red}} = \frac{1}{\sqrt{3}}\begin{pmatrix} 0 & 0 & 0 & 0 \\ 0 & \tau_1 & \sqrt{\tau_1 \tau_3} & 0 \\ 0 & \sqrt{\tau_1 \tau_3} & \tau_3 & 0 \\ 0 & 0 & 0 & \tau_2 \end{pmatrix} \tag{5.73}$$

$$\hat{\rho}_{\text{red}}^{T_2} = \frac{1}{\sqrt{3}}\begin{pmatrix} 0 & 0 & 0 & \sqrt{\tau_1 \tau_3} \\ 0 & \tau_1 & 0 & 0 \\ 0 & 0 & \tau_3 & 0 \\ \sqrt{\tau_1 \tau_3} & 0 & 0 & \tau_2 \end{pmatrix}. \tag{5.74}$$

The relevant negative eigenvalue reads $\lambda_- = \frac{1}{2}\left(\tau_2 - \sqrt{\tau_2^2 + 4\tau_1\tau_3}\right)$ and the logarithmic negativity becomes

$$\mathcal{L} = \ln\left(1 - \tau_2 + \sqrt{\tau_2^2 + 4\tau_1\tau_3}\right) \tag{5.75}$$

which matches the earlier result for the equal tripartition in the limit $\tau_k = 1/3$. This result underlines that negativity exhibits universality in this model, in that (5.75) differs markedly from (5.52) commensurate with the fact that we are dealing with a different universality class in this section.

References

1. H.J. Lipkin, N. Meshkov, A.J. Glick, Validity of many-body approximation methods for a solvable model. Nucl. Phys. **62**, 188 (1965)
2. N. Meshkov, A.J. Glick, H.J. Lipkin, Validity of many-body approximation methods for a solvable model. Nucl. Phys. **62**, 199 (1965)
3. A.J. Glick, H.J. Lipkin, N. Meshkov, Validity of many-body approximation methods for a solvable model. Nucl. Phys. **62**, 211 (1965)
4. R. Botet, R. Jullien, Large-size critical behavior of infinitely coordinated systems. Phys. Rev. B **28**, 3955 (1983)
5. J. Anders, Diploma thesis: Estimating the Degree of Entanglement of Unknown Gaussian States. (University of Potsdam, Germany, 2003)
6. A. Serafini, PhD Thesis: Decoherence and Entanglement in Continuous Variable Quantum Information. University of Salerno, Italy, 2004
7. M. Cramer, Quasi-Free Systems on General Lattices: Criticality, Entanglement-Area laws, and Single-Copy Entanglement. PhD thesis, University of Potsdam, Germany, 2006
8. R. Simon, Peres-horodecki separability criterion for continuous variable systems. Phys. Rev. Lett. **84**, 2726 (2000)
9. J.W. Negele, H. Orland, *Quantum Many-Particle systems*. (Addison Wesley, Reading, MA, 1988)
10. A.B. Dutta, N. Mukunda, R. Simon, The real symplectic groups in quantum mechanics and optics. Pramana, **45**, 471 (1995)
11. J. Williamson, On the normal forms of linear canonical transformations in dynamics. Am. J. Mathematics **58**, 58 (1936)
12. R. Simon, S. Chaturvedi, V. Srinivasan, Congruences and canonical forms for a positive matrix: Application to the schweinler–wigner extremum principle. J. Mathematical Phys. **40**(7), 3632–3642 (1999)
13. L.E. Ballentine, *Quantum Mechanics: A Modern Development*. (World Scientific Publishing Co. Pte. Ltd, Singapore, 1998)
14. S. Dusuel, J. Vidal, Unitary transformations and finite-size scaling exponents in the Lipkin–Meshkov–Glick model. Phys. Rev. B **71**, 224420 (2005)
15. T. Holstein, H. Primakoff, Field dependence of the intrinsic domain magnetization of a ferromagnet. Phys. Rev. **58**, 1098 (1940)
16. M. Cramer, J. Eisert, Correlations, spectral gap, and entanglement in harmonic quantum systems on generic lattices. New J. Phys. **8**, 71 (2006)
17. N. Schuch, J.I. Cirac, M.M. Wolf, Quantum states on harmonic lattices. Commun. Math. Phys. **267**, 65 (2006)

18. G. Vidal, R.F. Werner, Computable measure of entanglement. Phys. Rev. A **65**(3), 032314 (2002)
19. A. Serafini, F. Illuminati, S. De Siena, Symplectic invariants, entropic measures and correlations of Gaussian states. J. Phys. B **37**, 21 (2004)
20. H. Wichterich, J. Vidal, S. Bose, Universality of the negativity in the Lipkin–Meshkov–Glick model. Phys. Rev. A **81**, 032311 (2010)
21. J. Vidal, S. Dusuel, T. Barthel, Entanglement entropy in collective models. J. Stat. Mech.: Theory Exp. P01015 (2007)
22. T. Barthel, S. Dusuel, J. Vidal, Entanglement entropy beyond the free case. Phys. Rev. Lett. **97**, 220402 (2006)
23. R. Orús, S. Dusuel, J. Vidal, Equivalence of critical scaling laws for many-body entanglement in the Lipkin–Meshkov–Glick model. Phys. Rev. Lett. **101**, 025701 (2008)
24. J. Eisert, M. Cramer, M.B. Plenio, Colloquium: Area laws for the entanglement entropy. Rev. Mod. Phys. **82**(1), 277–306 (2010)
25. J.I. Latorre, R. Orús, E. Rico, J. Vidal, Entanglement entropy in the Lipkin–Meshkov–Glick model. Phys. Rev. A **71**, 064101 (2005)

Chapter 6
Conclusions and Outlook

In this treatise, we have shed some light on the subject of entanglement between noncomplementary regions of many-body system.

We have seen in Chap. 2 that the coherent dynamics after a quantum quench can give rise to a substantial amount of entanglement between a designated pair of spins in a spin chain. It seems feasible that this form of entanglement could be evidenced in an actual experiment. Future work could focus on more realistic quench scenarios, respecting the experimentally achievable limits within which parameters can be adjusted, as well as the finite temporal rate at which such a quench could be performed (our study assumed instantaneous quenches, so far).

The work of Chap. 3 focused on the question of whether a measurement on disjoint regions of many-body system can give rise to a pure and entangled state of these regions. We have seen by studying ground states of quantum spin chains that particular measurements, namely those where a measurement outcome is compatible with more than one quantum state of the regions, can accomplish this. The probabilistic aspects of measurement clearly imply that such a pure entangled state could not be achieved at will, but one would have to assume that the experimenter would be granted sufficient time to learn about the measurement outcome and that the state would not change in the mean time. Yet, the concept of localisable entanglement [1, 2] relies on similar such assumptions.

In the context of ground states, and particularly at quantum phase transitions, entanglement between noncomplementary regions of particles is still far from being fully understood. From a theoretical perspective, it would be interesting to find an explanation for the emergence of an exponential decay of entanglement that was evidenced in Chap. 4, and which seems surprising in view of the ubiquitous power-law decays in critical systems. A future direction could be to study this subject from the perspective of conformal field theory [3], where entanglement of disjoint regions (that with the remainder of the system but not between the regions) has recently attracted much interest [4, 5]. However, it is still unclear how one would quantify entanglement between the regions in this framework.

H.C. Wichterich, *Entanglement Between Noncomplementary Parts of Many-Body Systems*, Springer Theses, DOI: 10.1007/978-3-642-19342-2_6, © Springer-Verlag Berlin Heidelberg 2011

Finally, in Chap. 5 entanglement of noncomplementary parts of interacting spins on an infinitely connected graph, described by the Lipkin–Meshkov–Glick model, has been investigated. This simple model allowed a rather complete analytical treatment of negativity in the thermodynamic limit. Recently, the quantum phase transition of the Dicke model [6] has been observed experimentally in an atomic cloud which interacts with a optical cavity mode [7]. Due to its similarities to the Lipkin–Meshkov–Glick model, it will be interesting to extend our study of Chap. 5 to the Dicke model and assess the possibility of evidencing the entanglement in an experimental setup.

References

1. F. Verstraete, M. Popp, J.I. Cirac, Entanglement versus correlations in spin systems. Phys. Rev. Lett. **92**(2), 027901 (2004)
2. M. Popp, F. Verstraete, M.A. Martín-Delgado, J.I. Cirac, Localizable entanglement. Phys. Rev. A **71**(4), 042306 (2005)
3. M. Henkel, *Conformal Invariance and Critical Phenomena* (Springer, Berlin, 1999)
4. V. Alba, L. Tagliacozzo, P. Calabrese, Entanglement entropy of two disjoint blocks in critical ising models. Phys. Rev. B **81**(6), 060411 (2010)
5. M. Fagotti, P. Calabrese, Entanglement entropy of two disjoint blocks in xy chains, J. Stat. Mech. **2010**, P04016 (2010)
6. R.H. Dicke, Coherence in spontaneous radiation processes. Phys. Rev. **93**, 99 (1954)
7. K. Baumann, C. Guerlin, F. Brennecke, T. Esslinger, Dicke quantum phase transition with a superfluid gas in an optical cavity. Nature **464**, 1301–1306 (2010)

Appendix A

Diagonalisation of the XX Model

The Hamiltonian of the XX spin chain reads as

$$\hat{H} = \sum_{l=1}^{N-1} \frac{J}{2} \left(\hat{\sigma}_l^x \hat{\sigma}_{l+1}^x + \hat{\sigma}_l^y \hat{\sigma}_{l+1}^y \right), \tag{A.1}$$

or, in terms of spin raising and lowering operators, $\sigma^{\pm} = \frac{1}{2}(\hat{\sigma}^x \pm i\hat{\sigma}^y)$

$$\hat{H} = \sum_{l=1}^{N-1} J \left(\hat{\sigma}_l^+ \hat{\sigma}_{l+1}^- + \hat{\sigma}_l^- \hat{\sigma}_{l+1}^+ \right) \tag{A.2}$$

and assumes a quadratic form

$$\hat{H}(\Delta = 0) = J \sum_{l=1}^{N-1} \hat{c}_l^\dagger \hat{c}_{l+1} + \hat{c}_{l+1}^\dagger \hat{c}_l. \tag{A.3}$$

in Jordan Wigner fermions [Eq. 2.19]. This form can be diagonalised [1] by introducing new fermionic operators

$$\hat{\eta}_k^\dagger = \sum_l g_{k,l} \hat{c}_l^\dagger \tag{A.4}$$

with real coefficients $g_{k,l}$ which must satisfy the condition

$$\sum_l g_{k,l} g_{l,m} = \delta_{k,m} \tag{A.5}$$

so that the η_k be fermion operators and satisfy the CAR. The ansatz

$$\hat{H} = \sum_k \epsilon_k \hat{\eta}_k^\dagger \hat{\eta}_k + \text{const.} \tag{A.6}$$

$$\Leftrightarrow \quad [\hat{\eta}_k, \hat{H}] - \epsilon_k \hat{\eta}_k = 0 \tag{A.7}$$

leads, when (A.7) is combined with Eqs. A.3 and A.4, to the following set of linear equations for $g_{k,l}$

$$J g_{k,l-1} + J g_{k,l+1} - \epsilon_k g_{k,l} = 0 \tag{A.8}$$

which subject to the boundary condition $g_{k,l} = 0$ for $k, l \in \{0, N+1\}$ and the constraint Eq. A.5 is solved by

$$g_{k,l} = \sqrt{\frac{2}{N+1}} \sin(q_k l) \tag{A.9}$$

$$\epsilon_k = 2J \cos(q_k) \tag{A.10}$$

$$q_k = \frac{\pi k}{N+1} \tag{A.11}$$

whereby the ansatz Eq. A.6 is verified.

Appendix B

Factorisation of Fermionic Correlation Functions

We will show here, that fourth order correlation functions of fermion operators c_k factorise according to

$$\langle \hat{c}_n^\dagger \hat{c}_m^\dagger \hat{c}_k \hat{c}_l \rangle = \langle \hat{c}_n^\dagger \hat{c}_l \rangle \langle \hat{c}_m^\dagger \hat{c}_k \rangle - \langle \hat{c}_n^\dagger \hat{c}_k \rangle \langle \hat{c}_m^\dagger \hat{c}_l \rangle \tag{B.1}$$

if the expectation value is taken with respect to a density operator

$$\hat{\rho} = K e^{-\tilde{\tilde{H}}}, \tag{B.2}$$

where K ensures normalisation and

$$\tilde{\tilde{H}} = \sum_{i,j} \hat{c}_i^\dagger \mathbb{A}_{i,j} \hat{c}_j, \quad \mathbb{A} \in \mathbb{R}^{4 \times 4} \tag{B.3}$$

$$\tilde{\tilde{H}}^\dagger = \tilde{\tilde{H}}. \tag{B.4}$$

By virtue of canonical anticommution relations (CAR) among c_k

$$[\hat{c}_k^\dagger, \hat{c}_l]_+ = \delta_{k,l} \tag{B.5}$$

$$[\hat{c}_k, \hat{c}_l]_+ = [\hat{c}_k^\dagger, \hat{c}_l^\dagger]_+ = 0 \tag{B.6}$$

it is implied that $\mathbb{A}^T = \mathbb{A}$. Henceforth,

$$\langle \cdots \rangle \equiv \mathrm{Tr} \left[\cdots K e^{-\tilde{\tilde{H}}} \right]. \tag{B.7}$$

Equation B.1 is an instance of Wick's theorem in statistical mechanics, see e.g. [2]. In analogy to the procedure presented in Appendix A, the quadratic form Eq. B.3 is diagonalised by introducing new fermionic operators

$$\hat{\eta}_k = \sum_{l=1}^{4} g_{k,l} \hat{c}_l, \quad g_{k,l} \in \mathbb{R} \tag{B.8}$$

where coefficients $g_{k,l}$ are the solutions to

$$\mathbb{A}\,\mathbf{g}_k = \tilde{\epsilon}_k\,\mathbf{g}_k, \quad \mathbf{g}_k = (g_{k,1}, g_{k,2}, g_{k,3}, g_{k,4})^T \in \mathbb{R}^4 \tag{B.9}$$

under the constraint Eq. A.5. It follows that [1]

$$\hat{\tilde{H}} = \sum_k \tilde{\epsilon}_k \hat{\eta}_k^\dagger \hat{\eta}_k + \text{const.} \tag{B.10}$$

and furthermore

$$\hat{\eta}_k \hat{\rho} = e^{-\tilde{\epsilon}_k} \hat{\rho} \hat{\eta}_k \tag{B.11}$$

$$\hat{\eta}_k^\dagger \hat{\rho} = e^{\tilde{\epsilon}_k} \hat{\rho} \hat{\eta}_k^\dagger \tag{B.12}$$

where in the second line we made use of the formula for noncommuting operators $\hat{A}$ and $\hat{B}$

$$e^{-\alpha\hat{A}} \hat{B} e^{\alpha\hat{A}} = \hat{B} - \alpha[\hat{A}, \hat{B}] + \frac{\alpha^2}{2!}[\hat{A}, [\hat{A}, \hat{B}]] \mp \cdots.$$

Therefore, we deduce

$$\langle \hat{c}_n^\dagger \hat{c}_m^\dagger \hat{c}_k \hat{c}_l \rangle = \sum_{i,j,\mu,\nu} g_{n,i} g_{m,j} g_{k,\mu} g_{l,\nu} \langle \hat{\eta}_i^\dagger \hat{\eta}_j^\dagger \hat{\eta}_\mu \hat{\eta}_\nu \rangle \tag{B.13}$$

$$\langle \hat{\eta}_i^\dagger \hat{\eta}_j^\dagger \hat{\eta}_\mu \hat{\eta}_\nu \rangle = \delta_{j,\mu} \langle \hat{\eta}_i^\dagger \hat{\eta}_\nu \rangle - \langle \hat{\eta}_i^\dagger \hat{\eta}_\mu \hat{\eta}_j^\dagger \hat{\eta}_\nu \rangle \tag{B.14}$$

$$= \delta_{j,\mu} \langle \hat{\eta}_i^\dagger \hat{\eta}_\nu \rangle - \delta_{j,\nu} \langle \hat{\eta}_i^\dagger \hat{\eta}_\mu \rangle + \langle \hat{\eta}_i^\dagger \hat{\eta}_\mu \hat{\eta}_\nu \hat{\eta}_j^\dagger \rangle \tag{B.15}$$

$$\overset{(B.12)}{=} \delta_{j,\mu} \langle \hat{\eta}_i^\dagger \hat{\eta}_\nu \rangle - \delta_{j,\nu} \langle \hat{\eta}_i^\dagger \hat{\eta}_\mu \rangle + e^{\tilde{\epsilon}_j} \langle \hat{\eta}_j^\dagger \hat{\eta}_i^\dagger \hat{\eta}_\mu \hat{\eta}_\nu \rangle \tag{B.16}$$

$$= \delta_{j,\mu} \langle \hat{\eta}_i^\dagger \hat{\eta}_\nu \rangle - \delta_{j,\nu} \langle \hat{\eta}_i^\dagger \hat{\eta}_\mu \rangle - e^{\tilde{\epsilon}_j} \langle \hat{\eta}_i^\dagger \hat{\eta}_j^\dagger \hat{\eta}_\mu \hat{\eta}_\nu \rangle \tag{B.17}$$

$$\Leftrightarrow \langle \hat{\eta}_i^\dagger \hat{\eta}_j^\dagger \hat{\eta}_\mu \hat{\eta}_\nu \rangle = \frac{\delta_{j,\mu}}{1 + e^{\tilde{\epsilon}_j}} \langle \hat{\eta}_i^\dagger \hat{\eta}_\nu \rangle - \frac{\delta_{j,\nu}}{1 + e^{\tilde{\epsilon}_j}} \langle \hat{\eta}_i^\dagger \hat{\eta}_\mu \rangle \tag{B.18}$$

The elementary correlation function are of second order and read as

$$\langle \hat{\eta}_k^\dagger \hat{\eta}_l \rangle = \delta_{k,l} - \langle \hat{\eta}_l \hat{\eta}_k^\dagger \rangle \tag{B.19}$$

$$\overset{(B.12)}{=} \delta_{k,l} - e^{\tilde{\epsilon}_k} \langle \hat{\eta}_k^\dagger \hat{\eta}_l \rangle \tag{B.20}$$

$$\Leftrightarrow \langle \hat{\eta}_k^\dagger \hat{\eta}_l \rangle = \frac{\delta_{k,l}}{1 + e^{\tilde{\epsilon}_k}}. \tag{B.21}$$

In the first step, we additionally made use of the cyclic property of the trace
$\text{Tr}[\hat{A}\hat{B}] = \text{Tr}[\hat{B}\hat{A}]$. Combining Eqs. (B.21) and (B.18) and (B.1) gives

$$\langle \hat{c}_n^\dagger \hat{c}_m^\dagger \hat{c}_k \hat{c}_l \rangle = \sum_{i,j,\mu,\nu} g_{n,i} g_{m,j} g_{k,\mu} g_{l,\nu} \left(\langle \hat{\eta}_j^\dagger \hat{\eta}_\mu \rangle \langle \hat{\eta}_i^\dagger \hat{\eta}_\nu \rangle + \langle \hat{\eta}_j^\dagger \hat{\eta}_\nu \rangle \langle \hat{\eta}_i^\dagger \hat{\eta}_\mu \rangle \right) \tag{B.22}$$

$$= \langle \hat{c}_m^\dagger \hat{c}_k \rangle \langle \hat{c}_n^\dagger \hat{c}_l \rangle - \langle \hat{c}_m^\dagger \hat{c}_l \rangle \langle \hat{c}_n^\dagger \hat{c}_k \rangle \tag{B.23}$$

which is the desired factorisation formula. It holds for time dependent problems as
well, given that Heisenberg operators may be decomposed according to

$$\hat{c}_l(t) = \sum_k f_{k,l}(t) \hat{c}_k, \quad \sum_m f_{m,k}^*(t) f_{m,l}(t) = \delta_{k,l} \tag{B.24}$$

Appendix C

Time Dependence of the Reduced Density Operator Following Quench

We show in this Appendix, that the time dependence of matrix elements a, b and c of $\hat{\rho}_{1,N}$ [Eq. 2.18] can be rewritten in terms of two-point correlation functions of the Jordan Wigner fermions

$$\hat{c}_l^\dagger \equiv \left(\prod_{n=1}^{l-1} -\hat{\sigma}_l^z \right) \hat{\sigma}_l^+. \tag{C.1}$$

Matrix entry a reads as

$$a = \langle \hat{P}_1^\uparrow \hat{P}_N^\uparrow \rangle \tag{C.2}$$

$$= \mathrm{Tr}\left[\hat{\rho}(t) \hat{\sigma}_1^+ \hat{\sigma}_1^- \hat{\sigma}_N^+ \hat{\sigma}_N^- \right] \tag{C.3}$$

$$= \frac{1}{2}\left(\langle \mathcal{N}_1 | e^{i\hat{H}t} \hat{\sigma}_1^+ \hat{\sigma}_1^- \hat{\sigma}_N^+ \hat{\sigma}_N^- e^{-i\hat{H}t} | \mathcal{N}_1 \rangle + \langle \mathcal{N}_2 | e^{i\hat{H}t} \hat{\sigma}_1^+ \hat{\sigma}_1^- \hat{\sigma}_N^+ \hat{\sigma}_N^- e^{-i\hat{H}t} | \mathcal{N}_2 \rangle \right). \tag{C.4}$$

The appearance of products of four spin operators in a will lead to quartic terms (four-point correlators) in fermion operators as well. These can be reduced to two-point correlators with the help of Wick's theorem (Appendix B).

However, this can not be done straight away for expectation values of the form $\langle \cdots \rangle = \mathrm{Tr}\left[\hat{\rho}_{1,N}(t) \cdots \right]$: Recall that the quench is triggered by an instantaneous change in the anisotropy parameter $\Delta : \infty \to 0$. This can be rephrased by saying that the ground state of the Ising Hamiltonian becomes subjected to time evolution under the XX Hamiltonian. The Ising Hamiltonian is not quadratic, but quartic in Jordan Wigner fermions, and therefore we find that the initial state $\hat{\rho}_0 = \frac{1}{2}(|\mathcal{N}_1\rangle\langle\mathcal{N}_1| + |\mathcal{N}_2\rangle\langle\mathcal{N}_2|)$ of Eq. 2.8 is *not* an exponential of a quadratic form. But then, this is the condition for Wick's theorem to hold (Appendix B). It is, however, easily seen to hold for $|\mathcal{N}_1\rangle$ and $|\mathcal{N}_2\rangle^1$ and also for times $t > 0$, if the time evolution is generated by a quadratic Hamiltonian. This is the case here. Hence, we apply Wick's theorem separately on Schrodinger picture expectation values

$$\langle\cdots\rangle_1 \equiv \langle\mathcal{N}_1|e^{iHt}\cdots e^{-iHt}|\mathcal{N}_1\rangle$$

and

$$\langle\cdots\rangle_2 \equiv \langle\mathcal{N}_2|e^{iHt}\cdots e^{-iHt}|\mathcal{N}_2\rangle.$$

Denote $\hat{X} = \otimes_{k=1}^{N}\hat{\sigma}_k^x$ so that $|\mathcal{N}_1\rangle = \hat{X}|\mathcal{N}_2\rangle$ and $[\hat{X}, \hat{H}] = 0$, then Eq. C.4 becomes

$$a = \frac{1}{2}\left(\langle\hat{\sigma}_1^+\hat{\sigma}_1^-\hat{\sigma}_N^+\hat{\sigma}_N^-\rangle_1 + \langle\hat{X}\hat{\sigma}_1^+\hat{\sigma}_1^-\hat{\sigma}_N^+\hat{\sigma}_N^-\hat{X}\rangle_1\right) \tag{C.5}$$

$$= \frac{1}{2}\left(\langle\hat{\sigma}_1^+\hat{\sigma}_1^-\hat{\sigma}_N^+\hat{\sigma}_N^-\rangle_1 + \langle\hat{\sigma}_1^-\hat{\sigma}_1^+\hat{\sigma}_N^-\hat{\sigma}_N^+\rangle_1\right) \tag{C.6}$$

and in terms of fermion operators we find using $(-\hat{\sigma}_l^z)^2 = \mathbb{1}$

$$a = \frac{1}{2}\left(\langle\hat{c}_1^\dagger\hat{c}_1\hat{c}_N^\dagger\hat{c}_N\rangle_1 + \langle\hat{c}_1\hat{c}_1^\dagger\hat{c}_N\hat{c}_N^\dagger\rangle_1\right). \tag{C.7}$$

Invoking Wick's theorem (Appendix B) gives

$$a = \langle\hat{c}_1^\dagger\hat{c}_1\rangle_1\langle\hat{c}_N^\dagger\hat{c}_N\rangle_1 - \langle\hat{c}_1^\dagger\hat{c}_N\rangle_1\langle\hat{c}_N^\dagger\hat{c}_1\rangle_1 - \frac{1}{2}\left(\langle\hat{c}_1^\dagger\hat{c}_1\rangle_1 + \langle\hat{c}_N^\dagger\hat{c}_N\rangle_1 - 1\right). \tag{C.8}$$

The matrix entry b is then determined by $\mathrm{Tr}[\hat{\rho}_{1,N}] = 2a + 2b = 1$. It remains to determine c which is slightly more subtle because the *non-local string* of operators occurring in definition of the Jordan Wigner fermions Eq. C.1 does not cancel:

$$c = \frac{1}{2}\left(\langle\hat{\sigma}_1^-\hat{\sigma}_N^+\rangle_1 + \langle\hat{\sigma}_1^-\hat{\sigma}_N^+\rangle_2\right) \tag{C.9}$$

$$= \frac{1}{2}\left(\left\langle\hat{\sigma}_1^-\left(\otimes_{l=1}^{N-1} - \hat{\sigma}_l^z\right)\hat{c}_N^\dagger\right\rangle_1 + c.c.\right) \tag{C.10}$$

$$= \frac{1}{2}\left((-1)^{M+1}\langle\hat{c}_N^\dagger\hat{c}_1\rangle_1 + c.c.\right). \tag{C.11}$$

Here, M is the conserved number of spin up states in the dynamical Néel state

$$\left(\sum_{l=1}^{N}\hat{P}_l^\uparrow\right)e^{-iHt}|\mathcal{N}_1\rangle = Me^{-iHt}|\mathcal{N}_1\rangle,$$

i.e. $M = N/2$ for even N and $M = (N-1)/2$ for odd N.

---

[1] one could think of a simple quadratic Hamiltonian describing a staggered magnetic field, which would have one of $|\mathcal{N}_1\rangle, |\mathcal{N}_2\rangle$ as non-degenerate ground state

Appendix D

Density Matrix Renormalisation Group Algorithm

The DMRG algorithm in its original form [3] is a non-perturbative computational method aimed at finding an optimal approximation for the ground state wave function of a Hamiltonian describing particles in one spatial dimension with preferably short-range interaction. Here we discuss this procedure for the ground-state of a spin-1/2 chain with open boundary conditions and nearest-neighbour interactions. Any wave function of such a 1D arrangement of spin-1/2 can be brought to the form

$$|\psi\rangle = \sum_{\sigma_{[1\cdots N]}} \left(\sum_{\alpha_{[1\cdots N-1]}} \Lambda^{[1]\sigma_1}_{\alpha_1} \sqrt{w_{\alpha_1}} \Lambda^{[2]\sigma_2}_{\alpha_1 \alpha_2} \cdots \sqrt{w_{\alpha_{N-1}}} \Lambda^{[N]\sigma_N}_{\alpha_{N-1}} \right) |\sigma_1, \sigma_2, \cdots, \sigma_N\rangle \quad (D.1)$$

as shown in Sect. 4.3. Contracting this tensor representation with respect to all indices σ_m and all but one of the indices α_m, $(m \neq l)$ gives rise to the Schmidt decomposition at bond l

$$|\psi\rangle = \sum_{\alpha_l} \sqrt{w_{\alpha_l}} |w^L_{\alpha_l}\rangle \otimes |w^R_{\alpha_l}\rangle, \quad \alpha_l = 1 \cdots \chi_l. \quad (D.2)$$

The DMRG algorithm proceeds in two basic steps. These are (*i*) system growth and (*ii*) variational optimisation, which are also referred to as infinite and finite system DMRG, respectively [3]. Our target will be an optimal approximation to the wave function of Eq. D.1 for a spin chain of length N.

System growth. Starting with a spin chain of $n = 2$ spin-1/2, the Hamiltonian is constructed in the computational basis, giving rise to one two-body term $h_{1,2}$ and two one-body terms h_1 and h_2 in the computational basis of two spins

$$\hat{H} = \hat{h}_1 + \hat{h}_{1,2} + \hat{h}_2,$$

assuming open boundary condition. The ground state of this short lattice is obtained through exact diagonalisation, and the Schmidt decomposition with respect to a bisection at bond $l = 1$ can easily be found by means of singular value decomposition (see, e.g. [4])

$$|\psi\rangle = \sum_{\alpha_1} \sqrt{w_{\alpha_1}} |w_{\alpha_1}^L\rangle \otimes |w_{\alpha_1}^R\rangle.$$

The notation above entails a certain ambiguity, in that at this stage of the DMRG procedure $|w_{\alpha_1}^L\rangle$ and $|w_{\alpha_1}^R\rangle$ do not coincide with the Schmidt basis states of the wave function of the spin chain with target size $N > 2$ [Eq. D.2]. First, this slight inconsistency is intended to prevent a cluttered notation. Second, these Schmidt basis states will, at the final stage of DMRG, be identified with an optimal approximation to those of Eq. D.2, justifying the lack of notational rigour.

Now, two additional spins will be added to the present configuration. In pictorial terms, using the notation which was introduced in Sect. 4.3, we seek for a representation of the form

$$[L_1] \bullet \bullet [R_1].$$

In the computational basis, "$\bullet \bullet \bullet\bullet$", the Hamiltonian would read as (note that the spin formerly labelled 2 is now labelled 4)

$$\hat{H} = \hat{h}_1 + \hat{h}_{1,2} + \hat{h}_2 + \hat{h}_{2,3} + \hat{h}_3 + \hat{h}_{3,4} + \hat{h}_4.$$

Our aim is, however, to represent $\hat{H}$ in a compact basis of Schmidt vectors $|w_\alpha^L\rangle$ and $|w_\alpha^R\rangle$ which were found earlier. To this end, we introduce left and right *block* Hamiltonians, with matrix representation [compare Eqs. (4.11) and (4.12)]

$$\langle w_{\alpha_1}^L |\hat{H}_L| w_{\alpha_1'}^L \rangle = \sum_{\sigma_1, \sigma_1'} \Lambda_{\alpha_1}^{[1]\sigma_1} \left(\Lambda_{\alpha_1'}^{[1]\sigma_1'} \right)^* \langle \sigma_1'|h_1|\sigma_1 \rangle \qquad (D.3)$$

$$\langle w_{\alpha_3}^R |\hat{H}_R| w_{\alpha_3'}^R \rangle = \sum_{\sigma_4, \sigma_4'} \Lambda_{\alpha_3}^{[4]\sigma_4} \left(\Lambda_{\alpha_3'}^{[4]\sigma_4'} \right)^* \langle \sigma_4'|h_4|\sigma_4 \rangle. \qquad (D.4)$$

Here, the tensors Λ which yield the basis transformations were obtained in the course of the Schmidt decomposition corresponding to a ground state with $n = 2$. Therefore, these basis states are not strictly appropriate for left and right blocks of the present setting with $n = 4$. We stress that DMRG is a variational procedure which ultimately gives rise to an accurate approximation, while during system growth no rigorous accuracy is intended.

Left and right blocks (presently containing one spin each) are interacting via the coupling terms $h_{1,2}$ and $h_{3,4}$ to their adjacent sites, respectively. When referring to the desired representation, $[L_1] \bullet \bullet [R_1]$, we will rename these operators as follows

$$\langle w^L_{\alpha_1}, \sigma'_2 | \hat{H}_{L\bullet} | w^L_{\alpha'_1}, \sigma_2 \rangle = \sum_{\sigma_{[1,2]}, \sigma'_{[1,2]}} \Lambda^{[1]\sigma_1}_{\alpha_1} \left(\Lambda^{[1]\sigma'_1}_{\alpha'_1} \right)^* \langle \sigma'_1, \sigma'_2 | h_{1,2} | \sigma_1, \sigma_2 \rangle \qquad (D.5)$$

$$\langle w^R_{\alpha_3}, \sigma'_4 | \hat{H}_{\bullet R} | w^R_{\alpha'_3}, \sigma_4 \rangle = \sum_{\sigma_{[3,4]}, \sigma'_{[3,4]}} \Lambda^{[4]\sigma_4}_{\alpha_3} \left(\Lambda^{[4]\sigma'_4}_{\alpha'_3} \right)^* \langle \sigma'_3, \sigma'_4 | h_{3,4} | \sigma_3, \sigma_4 \rangle \qquad (D.6)$$

finally giving rise to the Hamiltonian of the $n = 4$ chain

$$\hat{H} = \hat{H}_L + \hat{H}_{L\bullet} + \hat{h}_2 + \hat{h}_{2,3} + \hat{h}_3 + \hat{H}_{\bullet R} + \hat{H}_R.$$

The growth procedure continues by iteratively carrying out the following steps, until the desired system size $n = N$ is reached

1. Find ground state of $\hat{H}$ by sparse diagonalisation (e.g., Lanczos method [5])
2. compute Schmidt decomposition with respect to the symmetric bisection, keeping only M vectors with the largest weights $\sqrt{w_\alpha}$
3. store all block operators and basis transformations of the present configuration
4. define new left and right blocks by adding a single site to the former blocks respectively
5. form operators $\hat{H}_L$, $\hat{H}_{L\bullet}$, $\hat{H}_{\bullet R}$ and $\hat{H}_R$ using the Schmidt bases from the previous step 2 and add two spins in between, thereby increasing the system size n by two.
6. form new Hamiltonian $\hat{H} = \hat{H}_L + \hat{H}_{L\bullet} + \hat{h}_{\frac{n}{2}} + \hat{h}_{\frac{n}{2}\frac{n}{2}+1} + \hat{h}_{\frac{n}{2}+1} + \hat{H}_{\bullet R} + \hat{H}_R$

Variational optimisation. Up to this point, the lattice was grown from an initial configuration of $n = 2$ spins by iteratively adding two sites. Now, that we have reached the desired system size $n = N$, our aim is to variationally optimise the tensors Λ and $\sqrt{w}$ which, at the present stage, constitute rather poor approximations to the actual tensors appearing in Eq. D.1.

This is achieved by successive so-called *finite-size sweeps*, which amount to gradually shifting the free lattice sites (initially placed at the bond $l = N/2$) by means of a concatenation of appropriate basis transformations to the left terminal sites, then to the right terminal sites, and back to the symmetric partition. Pictorially, in a middle-to-left sweep the right block is grown at the expense of the left block, thereby keeping N constant

$$[L_{N/2-1}] \bullet \bullet [R_{N/2-1}] \rightarrow [L_{N/2-2}] \bullet \bullet [R_{N/2}] \rightarrow \cdots \rightarrow \bullet \bullet [R_{N-2}].$$

Next, a left-to-right sweep grows the left block at the expense of the right block

$$\bullet \bullet [R_{N-2}] \rightarrow [L_1] \bullet \bullet [R_{N-3}] \rightarrow \cdots \rightarrow [L_{N-2}] \bullet \bullet,$$

and so forth. Consider, for example, a right-to-left sweep. Each individual change of representation (referring to all operator representations as well as the ground state representation) of the form

$$[L_{l-1}] \bullet \bullet [R_{N-l-1}] \rightarrow [L_{l-2}] \bullet \bullet [R_{N-l}],$$

is partly governed by the most recent tensor $\Lambda^{[l+1]\sigma_{l+1}}_{\alpha_l\alpha_{l+1}}$ which was obtained in the course of the Schmidt decomposition at bond l [compare Eqs. (4.6)–(4.10)]. This change of representation is instrumental in growing the right block Hamiltonian $\hat{H}_R$ by a single site. More explicitly, the operator $\hat{H}_R$ of the forthcoming representation is equivalent, up to a similarity transformation, to the operator $\hat{H}_R + \hat{H}_{\bullet R} + h_{l+1}$ of the present representation. By contrast, the forthcoming left block Hamiltonian $\hat{H}_L$, being of shrinking size, is read from the memory. It is therefore reminiscent of an earlier left-to-right sweep (or, where required, the system growth procedure) and will be updated in a subsequent sweep.[2] Operators $\hat{H}_{\bullet R}$ and $\hat{H}_{L\bullet}$ are constructed in basically the same vein. For an in-depth discussion, including details on several technical subtleties, we refer the reader to [6].

Subsequently, the ground state of $\hat{H}$ is found in each of these representations, using an appropriate sparse algorithm. Finally, the Schmidt decomposition (truncated at M-th order) is computed and the corresponding tensors Λ and $\sqrt{w}$ are determined and are used to replace stored tensors from previous sweeps.

In this spirit, the representation Eq. D.1 is approached with an accuracy depending on the system size N, the particulars of the ground state wave-function, and the chosen bond dimension M. After usually five to ten full sweeps the variational procedure has converged towards an optimal approximation of the ground state of

$$\hat{H} = \sum_{l=1}^{N-1}\left(\hat{h}_l + \hat{h}_{l,l+1}\right) + \hat{h}_N,$$

in terms of tensors Λ and $\sqrt{w}$ [see Eq. D.1].

---

[2] unless the variational optimisation terminates beforehand

Appendix E

Proof of Williamson's theorem

In what follows, we will present a simple proof of William's theorem 5.1.2 (for the case of real, symmetric positive definite matrices). The following property of the eigenspectrum of a product of matrices (a consequence of Sylvester's determinant theorem) will turn out to be useful in this context:

Lemma E.0.1 *(Cyclicity of characteristic polynomial) Let matrices X and Y be square matrices and let Y be non-singular. Then, the matrices XY and YX have the same eigenvalues.*

Proof Let w_m denote the eigenvalues of matrix XY, so they are the solutions to the characteristic polynomial

$$\det(XY - w_m \mathbb{1}) = 0. \tag{E.1}$$

One has that

$$\det(XY - w_m \mathbb{1}) = \det(Y^{-1}(YX - w_m \mathbb{1})Y) = \det(YX - w_m \mathbb{1}) \tag{E.2}$$

which proves the lemma. $\qquad\qquad\square$

Let us restate Williamsons theorem, for convenience,

Theorem E.0.2 (Williamson) *Let M be a $2N \times 2N$ real symmetric and positive definite matrix, then there exists a symplectic transformation $\mathbb{S} \in Sp(2N, \mathbb{R})$ such that*

$$\mathbb{S}M\mathbb{S}^T = \Lambda \oplus \Lambda \tag{E.3}$$

with a diagonal $N \times N$ positive definite diagonal matrix $\Lambda = \mathrm{diag}(\lambda_1, \lambda_2, \cdots, \lambda_N)$. The symplectic eigenvalues λ_n are given by the positive square roots of the eigenvalues of $(i\Omega M)^2$.

Proof (As found in Ref. [7]) The matrix which solves Eq. E.3 reads as

$$\mathbb{S} = DOM^{-1/2}$$

the factors of which being the $2N \times 2N$ matrices $M^{-1/2}$ (real symmetric positive definite), $O \Leftrightarrow O^T O = OO^T = \mathbb{1}_N$ (orthogonal), and $D = \Lambda^{1/2} \oplus \Lambda^{1/2}$ (diagonal and positive definite). For the so defined $\mathbb{S}$ to be element of $Sp(2N, \mathbb{R})$ we have to ask

$$\mathbb{S}\Omega\mathbb{S}^T = DOM^{-1/2}\Omega M^{-1/2}O^T D \overset{!}{=} \Omega \tag{E.4}$$

The matrix $M^{-1/2}\Omega M^{-1/2}$ is antisymmetric owing to $\Omega^T = -\Omega$ and nonsingular because its factors are nonsingular. Therefore, there exists an orthogonal O such that

$$OM^{-1/2}\Omega M^{-1/2}O^T = \begin{pmatrix} \mathbb{0}_N & -\Lambda^{-1} \\ \Lambda^{-1} & \mathbb{0}_N \end{pmatrix}, \tag{E.5}$$

with Λ^{-1} diagonal and positive definite. Hence, with this choice of O, $\mathbb{S}$ satisfies Eq. E.4 and therefore $\mathbb{S} \in Sp(2N, \mathbb{R})$ concluding the proof of the first part of the theorem.

The second part can be proved as follows. Denoting the eigenspectrum of a matrix Y by $\mathrm{spec}(Y)$, and making use of the cyclic property of $\mathrm{spec}(XY)$ (Y nonsingular, see Lemma E.0.1) as well as $\Omega^2 = -\mathbb{1}$, one finds

$$\mathrm{spec}((i\Omega M)^2) = \mathrm{spec}\left(-(\Omega\mathbb{S}^{-1}(\Lambda \oplus \Lambda)\mathbb{S}^{-T})^2\right) \tag{E.6}$$

$$= \mathrm{spec}\left(-(\Omega^2\mathbb{S}^T\Omega^T(\Lambda \oplus \Lambda)\mathbb{S}^{-T})^2\right) \tag{E.7}$$

$$= \mathrm{spec}\left(-(\Omega(\Lambda \oplus \Lambda))^2\right) \tag{E.8}$$

$$= \mathrm{spec}\left(\Lambda^2 \oplus \Lambda^2\right) \tag{E.9}$$

implying that the eigenvalues w_m^2 of $(i\Omega M)^2$ are related to the symplectic spectrum λ_m of M as $w_m^2 - \lambda_m^2 = 0$.   □

Appendix F

Partial Transposition in Continuous Variable Systems

In this section we set forth the correspondence between partial transposition in Hilbert space and partial time-reversal in phase space. Let $\hat{\rho} \in \mathcal{H}_A \otimes \mathcal{H}_B$ be the density operator of a composite continuous variable quantum system. Partial transposition $((T_A \otimes \mathbb{1}_B)\hat{\rho})$ can be seen to be equivalent to reversing the momenta of Alice's subsystem $(p_A \to -p_A)$ of the quasi-probability distribution describing the state. Making use of the operator identity $e^{X+Y} = e^X e^Y e^{-[X,Y]/2}$ which holds when the individual operators X and Y commute with their commutator, we may rewrite the characteristic function Eq. 5.9 as follows

$$\chi(-\xi) = \text{Tr}\left[e^{-i(\xi_p^T \hat{\mathbf{x}} - \xi_x^T \hat{\mathbf{p}})}\hat{\rho}\right] \tag{F.1}$$

$$= \text{Tr}\left[e^{i\xi_x^T \hat{\mathbf{p}}}e^{-i\xi_p^T \hat{\mathbf{x}}}\hat{\rho}\right]e^{i\xi_x^T \xi_p/2} \tag{F.2}$$

$$= \int d^N q \langle \mathbf{q}|e^{i\xi_x^T \hat{\mathbf{p}}/2}e^{-i\xi_p^T \hat{\mathbf{x}}}\hat{\rho}e^{i\xi_x^T \hat{\mathbf{p}}/2}|\mathbf{q}\rangle e^{i\xi_x^T \xi_p/2} \tag{F.3}$$

$$= \int d^N q e^{-i\xi_p^T \mathbf{q}}\langle \mathbf{q} - \xi_x/2|\hat{\rho}|\mathbf{q} + \xi_x/2\rangle, \tag{F.4}$$

where in Eq. F.3 we wrote the trace in coordinate representation and subsequently used that position displacements are generated by the momentum operator, $\exp(i\delta\hat{p})|x\rangle = |x + \delta\rangle$. The Wigner function Eq. 5.8 becomes

$$\mathcal{W}(\mathbf{x}, \mathbf{p}) = \frac{1}{(2\pi)^N}\int d^N q \int d^N \xi_x \int d^N \xi_p e^{i\xi_p^T(\mathbf{x}-\mathbf{q})}\langle \mathbf{q} - \xi_x/2|\hat{\rho}|\mathbf{q} + \xi_x/2\rangle e^{i\xi_x^T \mathbf{p}}$$

$$= \frac{1}{(2\pi)^N}\int d^N \xi_x \langle \mathbf{x} - \xi_x/2|\hat{\rho}|\mathbf{x} + \xi_x/2\rangle e^{-i\xi_x^T \mathbf{p}}. \tag{F.5}$$

We see that the Wigner function is a Fourier transform of certain matrix elements of $\hat{\rho}$, here in coordinate representation. The canonical degrees of freedom can be regarded as being subdivided into those belonging to Alice and those belonging to Bob, i.e. $(\mathbf{x}, \mathbf{p}) : (\mathbf{x_A}, \mathbf{x_B}, \mathbf{p_A}, \mathbf{p_B})$. Partial transposition $(T_A \otimes \mathbb{1}_B)\hat{\rho}$ then amounts to reversing those momenta within the Wigner function belonging to Alice, $\mathcal{W}(\mathbf{x_A}, \mathbf{x_B}, \mathbf{p_A}, \mathbf{p_B}) \rightarrow \mathcal{W}(\mathbf{x_A}, \mathbf{x_B}, -\mathbf{p_A}, \mathbf{p_B})$, as can be directly inferred from Eq. F.5 by changing the sign of the corresponding integration variables $(\xi_x)_A$. The usefulness of this correspondence becomes particularly apparent for the computation of bipartite entanglement measures which are based on partial transposition and when further $\hat{\rho}$ belongs to the class of Gaussian states. This will be recalled in Appendix I.

Appendix G

Gaussian Wigner Representation of Bosonic Vacuum

Let bosonic operators be defined in the usual way $\hat{a}_k = (\hat{x}_k + i\hat{p}_k)/\sqrt{2}$. The bosonic vacuum, defined as $\hat{a}_k|0\rangle = 0 \; \forall \; k$, is an example for a Gaussian state as will be shown now. In position representation we have

$$0 = \frac{1}{\sqrt{2}}\langle\mathbf{x}|(\hat{x}_k + i\hat{p}_k)|0\rangle \tag{G.1}$$

$$= \frac{1}{\sqrt{2}}(x_k + i\partial_{x_k})\langle\mathbf{x}|0\rangle \tag{G.2}$$

$$\phi_0(\mathbf{x}) \equiv \langle x|0\rangle, \qquad \int d^N q\,\phi_0^*(\mathbf{q})\phi_0(\mathbf{q}) = 1 \tag{G.3}$$

$$\Leftrightarrow \phi_0(\mathbf{x}) = \pi^{-N/4}\exp\left(-\frac{1}{2}\sum_k x_k^2\right). \tag{G.4}$$

The characteristic function (F.4) assumes

$$\chi(\xi) = \int d^N q \; e^{i\xi_p^T \mathbf{q}}\phi_0(\mathbf{q} + \xi_x/2)\phi_0^*(\mathbf{q} - \xi_x/2) \tag{G.5}$$

$$= \pi^{-N/2}\int d^N q \; e^{i\xi_p^T \mathbf{q}}e^{-\frac{1}{2}(\mathbf{q}+\xi_x/2)^T(\mathbf{q}+\xi_x/2)}e^{-\frac{1}{2}(\mathbf{q}-\xi_x/2)^T(\mathbf{q}-\xi_x/2)} \tag{G.6}$$

$$= \pi^{-N/2}\,e^{-\xi_x^T\xi_x/4}\int d^N q \; e^{-\mathbf{q}^T\mathbf{q}+i\xi_p^T\mathbf{q}} \tag{G.7}$$

$$= e^{-\xi^T\xi/4}. \tag{G.8}$$

In the last line we invoked the identity for Gaussian integration [8]

$$\frac{1}{\left(\sqrt{2\pi}\right)^N} \int d^N q \, e^{-\frac{1}{2}\mathbf{q}^T \gamma \mathbf{q} + i\mathbf{x}^T \mathbf{q}} = \frac{1}{\sqrt{\det \gamma}} e^{-\frac{1}{2}\mathbf{x}^T (\gamma^{-1})\mathbf{x}}, \qquad (G.9)$$

where γ is a real symmetric positive $N \times N$ matrix and $\mathbf{x} \in \mathbb{R}^N$. By inspection of (G.8) it is revealed that the bosonic vacuum is a Gaussian state, see definition 5.1.3, with covariance matrix $\Gamma = \mathbb{1}$ and vanishing first moments $d_i = 0$.

Appendix H

Ground State Covariance Matrix of a Quadratic Hamiltonian

We show in this section that the covariance matrix of the ground state of a Hamiltonian which is quadratic in bosonic operators

$$\hat{H} = \sum_{k,l=1}^{N} \hat{a}_k^\dagger \mathbb{A}_{k,l} \hat{a}_l + \frac{1}{2}(\hat{a}_k^\dagger \mathbb{B}_{k,l} \hat{a}_l^\dagger + \text{H.c.}) \tag{H.1}$$

can be calculated in terms of the adjacency matrices $\mathbb{A}, \mathbb{B} \in \mathbb{R}^{N \times N}$ (which we assume to be real, symmetric, and positive). We start with diagonalising $\hat{H}$. To this end, it will be convenient to cast $\hat{H}$ into matrix form

$$\hat{H} = \frac{1}{2}\begin{pmatrix} \hat{\mathbf{a}} \\ \hat{\mathbf{a}}^\dagger \end{pmatrix}^T \begin{pmatrix} \mathbb{B} & \mathbb{A} \\ \mathbb{A} & \mathbb{B} \end{pmatrix} \begin{pmatrix} \hat{\mathbf{a}} \\ \hat{\mathbf{a}}^\dagger \end{pmatrix} \tag{H.2}$$

which by using Eq. 5.3 turns into

$$\hat{H} = \frac{1}{2}\begin{pmatrix} \hat{\mathbf{x}} \\ \hat{\mathbf{p}} \end{pmatrix}^T \begin{pmatrix} V_x & 0 \\ 0 & V_p \end{pmatrix} \begin{pmatrix} \hat{\mathbf{x}} \\ \hat{\mathbf{p}} \end{pmatrix}, \tag{H.3}$$

$$V_x = \mathbb{A} + \mathbb{B} \tag{H.4}$$

$$V_p = \mathbb{A} - \mathbb{B}. \tag{H.5}$$

Following loosely [9], we can diagonalise $\hat{H}$ by means of a concatenation of two symplectic transformations. The first, $\mathbb{S}_1 = (V_x^{-1/2}) \oplus (V_x^{1/2}) \Leftrightarrow \hat{\mathbf{r}} = \mathbb{S}_1 \hat{\mathbf{r}}'$, turns (H.3) into

$$\hat{H} = \frac{1}{2}(\hat{\mathbf{r}}')^T \begin{pmatrix} \mathbb{1} & 0 \\ 0 & V_x^{1/2} V_p V_x^{1/2} \end{pmatrix} \hat{\mathbf{r}}' \tag{H.6}$$

while the second, $\mathbb{S}_2 = OD^{1/2} \oplus OD^{-1/2}$, involves an orthogonal transformation $O^T O = OO^T = \mathbb{1}$ and a diagonal matrix $D = \mathrm{diag}(\epsilon_1, \epsilon_2, \ldots, \epsilon_N)$ so that $V_x^{1/2} V_p V_x^{1/2} = OD^2 O^T$. Finally, with $\hat{\mathbf{r}}' = \mathbb{S}_2 \hat{\mathbf{r}}''$, the Hamiltonian assumes its diagonal form

$$\hat{H} = \frac{1}{2}(\hat{\mathbf{r}}'')^T \begin{pmatrix} D & 0 \\ 0 & D \end{pmatrix} \hat{\mathbf{r}}'' = \sum_{m=1}^{N} \epsilon_m \hat{\eta}_m^\dagger \hat{\eta}_m + \mathrm{const.} \qquad (H.7)$$

where bosonic operators η_m and canonical operators $\hat{\mathbf{r}}''$ are connected through Eq. 5.3. By virtue of Eq. 5.15 and $\Gamma'' = \mathbb{1}$ the ground state covariance matrix in terms of original degrees of freedom $\hat{\mathbf{r}} = \mathbb{S}_1 \mathbb{S}_2 \hat{\mathbf{r}}''$ is given by $\Gamma = \mathbb{S}\mathbb{S}^T$ where $\mathbb{S} = \mathbb{S}_1 \mathbb{S}_2$. Therefore, using $\mathbb{S}_1^T = \mathbb{S}_1$ we obtain the final result [9, 10]

$$\Gamma = \mathbb{S}_1 (\mathbb{S}_2 \mathbb{S}_2^T) \mathbb{S}_1 \qquad (H.8)$$

$$= (V_x^{-1/2} \oplus V_x^{1/2})(ODO^T \oplus OD^{-1}O^T)(V_x^{-1/2} \oplus V_x^{1/2}) \qquad (H.9)$$

$$= V_x^{-1/2}(V_x^{1/2} V_p V_x^{1/2})^{1/2} V_x^{-1/2} \oplus V_x^{1/2}(V_x^{1/2} V_p V_x^{1/2})^{-1/2} V_x^{1/2}. \qquad (H.10)$$

One could have, alternatively, concatenated the two symplectic transformations $\mathbb{S}_3 = (V_p^{1/2}) \oplus (V_p^{-1/2})$ and $\mathbb{S}_4 = O_2 D^{1/2} \oplus O_2 D^{-1/2}$ with $V_p^{1/2} V_x V_p^{1/2} = O_2 D^2 O_2^T$. This diagonalises $\hat{H}$ as well, and the resulting form of the covariance matrix is

$$\Gamma = \mathbb{S}_3 (\mathbb{S}_4 \mathbb{S}_4^T) \mathbb{S}_3 \qquad (H.11)$$

$$= V_p^{1/2}(V_p^{1/2} V_x V_p^{1/2})^{-1/2} V_p^{1/2} \oplus V_p^{-1/2}(V_p^{1/2} V_x V_p^{1/2})^{1/2} V_p^{-1/2}. \qquad (H.12)$$

Often one representation turns out to be easier to evaluate than the other, for instance if one of V_x, V_p is diagonal.

Appendix I

Bipartite Entanglement of Gaussian States

We saw that in Sect. 1.1 that $\|\hat{\rho}^{T_A}\| > 1$, $\hat{\rho} \in \mathcal{H}_A \otimes \mathcal{H}_B$ is an important indicator of bipartite entanglement in mixed quantum states, for it can be used to provide a lower bound on entanglement of formation [11], which in turn can be given operational interpretation in terms of entanglement cost. We will now show that if $\hat{\rho}$ is a Gaussian state (or, more generally, an operator with Gaussian characteristic function), then $\|\hat{\rho}^{T_A}\|$ can be evaluated efficiently. To this end, we make use of the property $\|X \otimes Y\| = \|X\|\|Y\|$ of the trace norm, and the following.

Lemma I.0.3 (Direct product of Gaussian states) *An N-fold direct product*

$$\hat{\rho} = \bigotimes_{m=1}^{N} \hat{\rho}_m \tag{I.1}$$

of Gaussian states $\hat{\rho}_m$ with individual covariance matrices Γ_m is itself a Gaussian state with covariance matrix

$$\Gamma = \bigoplus_{m=1}^{N} \Gamma_m \tag{I.2}$$

in canonical coordinates $\hat{\mathbf{r}} = \oplus_m \hat{\mathbf{r}}_m$.
Proof From Eq. F.5 one has that for a general $\hat{\rho} = \otimes_{m=1} \hat{\rho}_m$

$$\mathcal{W}(\mathbf{x}, \mathbf{p}) = \prod_m \mathcal{W}_m(\mathbf{x}_m, \mathbf{p}_m) \tag{I.3}$$

by inspection. Let now $\hat{\rho}$ be a Gaussian state and let $\chi(\xi) = \exp(-(\Omega\xi)^T \Gamma \Omega \xi / 4)$ be its characteristic function (the first moments can always be made to vanish by appropriate phase-space translations), then with (G.9)

$$\mathcal{W}(\mathbf{x}, \mathbf{p}) = \frac{1}{(2\pi)^N} \int d^{2N}\xi \, e^{-\frac{1}{4}(\Omega\xi)^T \Gamma \Omega\xi} e^{i\xi^T \Omega \mathbf{r}} = \frac{e^{-\mathbf{r}^T(\Gamma^{-1})\mathbf{r}}}{\sqrt{\det(\Gamma/2)}}. \tag{I.4}$$

With $\Gamma = \oplus_n \Gamma_n$ and the properties

$$(\oplus_m \Gamma_m)^{-1} = \oplus_m \Gamma_m^{-1}$$

and

$$\det(\oplus_m \Gamma_m) = \prod_m \det \Gamma_m$$

we find

$$\mathcal{W}(\mathbf{x}, \mathbf{p}) = \prod_m \frac{e^{-\mathbf{r}_m^T(\Gamma_m^{-1})\mathbf{r}_m}}{\sqrt{\det(\Gamma_m/2)}}. \tag{I.5}$$

Hence, the Wigner functions $\mathcal{W}_m$ of Eq. I.3 are themselves of Gaussian form which concludes the proof   □

For a Gaussian state a consequence of the above Lemma in conjunction with William's theorem is that, in canonical variables $\hat{\mathbf{r}} = (\hat{x}_1, \hat{p}_1, \hat{x}_2, \hat{p}_2, \dots, \hat{x}_N, \hat{p}_N)$ which bring the covariance matrix to normal form,

$$\Gamma = \bigoplus_{m=1}^{N} \Gamma_m, \qquad \Gamma_m = \text{diag}(\lambda_m, \lambda_m)$$

$\hat{\rho}$ becomes a direct product $\hat{\rho} = \otimes_m \hat{\rho}_m$, $m = 1 \cdots N$ of Gaussian states $\hat{\rho}_m$ with covariance matrices $\Gamma_m = \text{diag}(\lambda_m, \lambda_m)$ and hence

$$\|\hat{\rho}\| = \prod_m \|\hat{\rho}_m\|. \tag{I.6}$$

Now, in order to compute $\|\hat{\rho}_m\|$ we will relate the eigenvalues $w_n^{(m)}$, $n = 0, 1, 2, \dots$ of $\hat{\rho}_m$ to the (two fold degenerate) symplectic eigenvalue λ_m of Γ_m. This is done in Appendix (J), and the result reads as [see Eq. J.12]

$$w_n^{(m)} = \frac{2}{\lambda_m + 1}\left[\frac{\lambda_m - 1}{\lambda_m + 1}\right]^n. \tag{I.7}$$

We finally obtain for the trace norm of the Gaussian state $\hat{\rho}$

$$\|\hat{\rho}\| = \prod_{m=1}^{N} \|\hat{\rho}_m\| = \prod_{m=1}^{N} \sum_{n=0}^{\infty} \frac{2}{\lambda_m + 1}\left|\frac{\lambda_m - 1}{\lambda_m + 1}\right|^n \tag{I.8}$$

$$= \prod_{m=1}^{N} \frac{2}{\lambda_m + 1 - |\lambda_m - 1|} = \begin{cases} 1 & \text{for } \lambda_m \geq 1, \\ \prod_{m=1}^{N} \frac{1}{\lambda_m} & \text{for } \lambda_m < 1. \end{cases} \tag{I.9}$$

The entire calculation of the trace norm $\|\hat{\rho}\|$ can be carried out analogously if the partial transpose $\hat{\rho}^{T_A}$ takes the role of $\hat{\rho}$, where only we have to replace Γ with Γ^{T_A} which is obtained from Γ by reversing all momenta of subsystem A (see Appendix F), implying that logarithmic negativity in terms of the symplectic spectrum λ_m of Γ^{T_A} reads as [12]

$$\mathcal{L} = \ln \|\hat{\rho}^{T_A}\| = -\sum_{\lambda_m < 1} \ln \lambda_m. \tag{I.10}$$

Appendix J

Density Matrix Spectra of Bosonic Gaussian States

Our aim in this section is to show that a Gaussian state $\hat{\rho}$ (assuming as usual vanishing first moments) with covariance matrix $\Gamma = \Gamma_x \oplus \Gamma_p$ is an exponential of a quadratic form, and that its spectrum can be related to the symplectic eigenvalues of its covariance matrix by following the methodology of [13]. To this end, we make the ansatz

$$\hat{\rho} = Ke^{-\sum_{kl}\hat{x}_k X_{k,l}\hat{x}_l + \hat{p}_k P_{k,l}\hat{p}_l} = Ke^{-\sum_m \tilde{\epsilon}_m(\hat{\eta}_m^\dagger \hat{\eta}_m + \frac{1}{2})}, \tag{J.1}$$

where K is a normalisation constant, X and P are real symmetric positive definite $N \times N$ matrices, and in the second step we diagonalised the quadratic form in the exponent along the very same lines presented in Appendix H, i.e. by means of a suitable symplectic transformation $\hat{\mathbf{r}} = (\mathbb{S}_x \oplus \mathbb{S}_p)\hat{\mathbf{r}}'$ leading to the diagonal form in bosonic operators $\hat{\eta}_m = (\hat{x}'_m + i\hat{p}_m)/\sqrt{2}$. We have that

$$(\Gamma_x)_{i,j} = 2\text{Tr}[\hat{x}_i \hat{x}_j \hat{\rho}] \tag{J.2}$$

$$= \sum_{m,n}(\mathbb{S}_x)_{m,i}(\mathbb{S}_x)_{n,j}\text{Tr}[(\hat{\eta}_m^\dagger + \hat{\eta}_m)(\hat{\eta}_n + \hat{\eta}_n^\dagger)\hat{\rho}] \tag{J.3}$$

and by using the formula for noncommuting operators A and B

$$e^{-\alpha A}Be^{\alpha A} = B - \alpha[A, B] + \frac{\alpha^2}{2!}[A, [A, B]] \mp \cdots$$

one obtains the relations

$$\hat{\eta}_m \hat{\rho} = e^{-\tilde{\epsilon}_m}\hat{\rho}\hat{\eta}_m \tag{J.4}$$

$$\hat{\eta}_m^\dagger \hat{\rho} = e^{+\tilde{\epsilon}_m}\hat{\rho}\hat{\eta}_m \tag{J.5}$$

which together with cyclic invariance of the trace lead to

$$(\Gamma_x)_{i,j} = \sum_{m,n} (\mathbb{S}_x)_{m,i} (\mathbb{S}_x)_{n,j} \left(\langle \hat{\eta}_m^\dagger \hat{\eta}_n \rangle + \langle \hat{\eta}_m \hat{\eta}_n^\dagger \rangle \right) \tag{J.6}$$

$$= \sum_m (\mathbb{S}_x)_{m,i} (\mathbb{S}_x)_{m,j} \left(\frac{1}{e^{\tilde{\epsilon}_m} - 1} + \frac{1}{1 - e^{-\tilde{\epsilon}_m}} \right) \tag{J.7}$$

$$(\Gamma_x)_{i,j} = \sum_m (\mathbb{S}_x)_{m,i} (\mathbb{S}_x)_{m,j} \coth \left(\frac{\tilde{\epsilon}_m}{2} \right) \tag{J.8}$$

It follows, that the pseudo-energies $\tilde{\epsilon}_m$ and the symplectic spectrum of Γ are related through

$$\lambda_m = \coth \left(\frac{\tilde{\epsilon}_m}{2} \right) \tag{J.9}$$

meaning the ansatz in (J.1) is verified, for it recovers all second moments of the state. It also leads to the correct factorisation of higher moments according to Wick's theorem [2] (Appendix B). The state $\hat{\rho}$ factorises in the Fock basis $\hat{\eta}_m^\dagger \hat{\eta}_m |n_m\rangle = n_m |n_m\rangle$, and by absorbing constant terms into K we have

$$\hat{\rho} = K' \bigotimes_{m=1}^{N} \left[\sum_{n_m=0}^{\infty} e^{-n_m \tilde{\epsilon}_m} |n_m\rangle\langle n_m| \right] \tag{J.10}$$

requiring $\mathrm{Tr}[\hat{\rho}] = 1$ and using the identity [14]

$$\coth^{-1}(x) = \frac{1}{2} \ln \left(\frac{x+1}{x-1} \right), \quad x \neq 0, 1$$

gives

$$\hat{\rho} = \bigotimes_{m=1}^{N} \left[\sum_{n_m=0}^{\infty} \frac{2}{\lambda_m + 1} \left(\frac{\lambda_m - 1}{\lambda_m + 1} \right)^{n_m} |n_m\rangle\langle n_m| \right] \tag{J.11}$$

which shows that the spectrum $w_n^{(m)}$, $n = 0, 1, 2, \ldots$, of the single mode density matrices $\hat{\rho}_m \Leftrightarrow \hat{\rho} = \otimes_m \hat{\rho}_m$ is related to the symplectic spectrum λ_m of Γ as

$$w_n^{(m)} = \frac{2}{\lambda_m + 1} \left(\frac{\lambda_m - 1}{\lambda_m + 1} \right)^n . \tag{J.12}$$

Appendix K

Bosonisation of the LMG Hamiltonian

Here, we derive the three mode boson representation of the Lipkin-Meshkov-Glick Hamiltonian which was stated in Eq. 5.28. We begin the derivation for the symmetric phase $h \geq 1$. Here, $\langle \hat{S}_x \rangle = \langle \hat{S}_y \rangle = 0$ such that the expectation value of the vector-valued ground state magnetisation $\mathbf{M} \sim \langle \hat{\mathbf{S}} \rangle$ points in z-direction. This is due to the spin-flip symmetry $[\hat{H}, \prod_k \hat{\sigma}_k^z] = 0$, which the unique ground state respects for $h \geq 1$. We depart from the following representation of the LMG Hamiltonian

$$\hat{H} = -\frac{1}{2N}(1+\gamma)\left(\hat{\mathbf{S}}^2 - \hat{S}_z^2\right) - \frac{1}{4N}(1-\gamma)(\hat{S}_+^2 + \hat{S}_-^2) - h\hat{S}_z. \tag{K.1}$$

In subsequent steps we will omit scalar valued terms as they will not be important for the discussion of ground state entanglement but only shift the energy spectrum. Since we are interested in the ground state properties, we are entitled to replace the operator $\hat{\mathbf{S}}^2$ by its expectation value $(N/2)(N/2+1)$. Then

$$\hat{H} \sim \frac{1}{2N}(1+\gamma)\hat{S}_z^2 - \frac{1}{4N}(1-\gamma)(\hat{S}_+^2 + \hat{S}_-^2) - h\hat{S}_z. \tag{K.2}$$

Using the mapping Eqs. 5.26 and 5.27 for $S = N/2$, one has that

$$\hat{S}_z = \sum_{k=1}^{3} \hat{S}_z^{(k)} = N/2 - \sum_{k=1}^{3} \hat{a}_k^\dagger \hat{a}_k \tag{K.3}$$

$$\hat{S}_z^2 = N^2/4 - N \sum_{k=1}^{3} \hat{a}_k^\dagger \hat{a}_k + \mathcal{O}(N^0) \tag{K.4}$$

and since $S_+ = \sqrt{N - \hat{a}^\dagger \hat{a}} \, \hat{a} = \sqrt{N}\hat{a} + \mathcal{O}(N^{-1/2})$

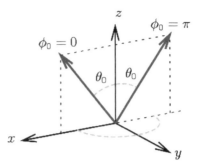

Fig. K.1 The two degenerate mean-field ground states. Figure reprinted from [15]

$$\hat{S}_+^2 = (\hat{S}_-^2)^\dagger = \left(\sum_{k=1}^{3} \hat{S}_+^{(k)}\right)^2 = \left(\sum_{k=1}^{3} \sqrt{N_k} a_k\right)^2 + \mathcal{O}(N^0). \tag{K.5}$$

Combining Eqs. K.2, and K.3–K.5 then leads to the quadratic representation of Eq. 5.28 for the symmetric phase $h \geq 1$. Now consider the symmetry-broken phase $0 \leq h < 1$, where a two-fold degeneracy develops in the thermodynamic limit [16], and the magnetisation of neither of the symmetry-broken ground states points into the spin-z direction [see Fig. K.1]. Therefore, the first thing to do is to rotate the spin operators so as to align the semi-classical magnetisation $\mathbf{M} = (N/2)(\sin\theta\cos\phi, \sin\theta\sin\phi, \cos\theta)$ ($\phi = 0$ or π) with the spin-z direction [15]

$$\begin{pmatrix} \hat{S}_x \\ \hat{S}_y \\ \hat{S}_z \end{pmatrix} = \begin{pmatrix} \cos\theta & 0 & \sin\theta \\ 0 & 1 & 0 \\ -\sin\theta & 0 & \cos\theta \end{pmatrix} \begin{pmatrix} \hat{S}_x' \\ \hat{S}_y' \\ \hat{S}_z' \end{pmatrix} \tag{K.6}$$

where $\theta = \arccos h$. In terms of the primed operators, the LMG Hamiltonian Eq. 5.19 becomes

$$\hat{H} = -\frac{1}{N}\left(\left(h\hat{S}_x' + \sqrt{1-h^2}\hat{S}_z'\right)^2 + \gamma\hat{S}_y'^2\right) - h^2\hat{S}_z' + h\sqrt{1-h^2}\hat{S}_x' \tag{K.7}$$

$$= -\frac{1}{N}\left[(1-h^2)\hat{S}_z'^2 + \frac{h^2+\gamma}{2}\left(\hat{S}_x'^2 + \hat{S}_y'^2\right) + \frac{h^2-\gamma}{2}\left(\hat{S}_x'^2 - \hat{S}_y'^2\right)\right]$$

$$- \frac{h\sqrt{1-h^2}}{N}(\hat{S}_x'\hat{S}_z' + \hat{S}_z'\hat{S}_x') - h^2\hat{S}_z' + h\sqrt{1-h^2}\hat{S}_x' \tag{K.8}$$

$$\hat{H} = -\frac{1}{N}\left[(1-h^2)\hat{S}_z'^2 + \frac{h^2+\gamma}{2}\left(\mathbf{S}'^2 - \hat{S}_z'^2\right) + \frac{h^2-\gamma}{4}\left(\hat{S}'^2_+ + \hat{S}'^2_-\right)\right]$$

$$- \frac{h\sqrt{1-h^2}}{2N}(\hat{S}_+'\hat{S}_z' + \hat{S}_z'\hat{S}_+' + H.c.) - h^2\hat{S}_z' + \frac{h\sqrt{1-h^2}}{2}(\hat{S}_+' + \hat{S}_-'). \tag{K.9}$$

Like before, we replace $\hat{S}'^2 \to (N/2)(N/2 + 1)$ and ignore the scalar valued terms, for they do not enter the ground state wave-function

$$\hat{H} \sim -\frac{1}{N}\left[\left(\frac{2 - \gamma - 3h^2}{2}\right)\hat{S}'^2_z + \frac{h^2 - \gamma}{4}(\hat{S}'^2_+ + \hat{S}'^2_-)\right]$$

$$-\frac{h\sqrt{1 - h^2}}{2N}(\hat{S}'_+\hat{S}'_z + \hat{S}'_z\hat{S}'_+ + H.c.) - h^2\hat{S}'_z + \frac{h\sqrt{1 - h^2}}{2}(\hat{S}'_+ + \hat{S}'_-). \quad (K.10)$$

By expanding $\hat{H}$ to order $\mathcal{O}((1/N)^0)$ using Eqs. K.3–K.5 we get[3]

$$\hat{H} \sim \frac{2 - \gamma - h^2}{2}\sum_k \hat{a}^\dagger_k\hat{a}_k + \frac{\gamma - h^2}{4N}\sum_{kl}\sqrt{N_kN_l}\left(\hat{a}^\dagger_k\hat{a}^\dagger_l + H.c.\right) + \mathcal{O}(N^{-1}) \quad (K.11)$$

which is the desired quadratic form of Eq. 5.28.

References

1. E. Lieb, T. Schultz, D. Mattis, Two soluble models of an antiferromagnetic chain. *Annals of Physics.* **16**(3), 407–466 (1961)
2. M. Gaudin, *Nuclear Physics.* **15**, 89 (1960)
3. S.R. White, *Phys. Rev. Lett.* **69**, 2863 (1992)
4. H.-P. Breuer, F. Petruccione, The theory of open quantum systems. (Clarendon Press, Oxford, 2002)
5. G.H. Golub, G.H. van Loan, Matrix computations. 3rd ed. (John Hopkins University Press, Baltimore, 1996)
6. A.L. Malvezzi, An introduction to numerical methods in low-dimensional quantum systems. *Braz. J. Phys.* **33**(1), (2003)
7. R. Simon, S. Chaturvedi, V. Srinivasan, Congruences and canonical forms for a positive matrix: Application to the schweinler–wigner extremum principle. *J. Math. phys.* **40**(7), 3632–3642 (1999)
8. J.W. Negele, H. Orland, Quantum many-particle systems. (Addison Wesley, Reading, MA, 1988)
9. M. Cramer, J. Eisert, Correlations, spectral gap, and entanglement in harmonic quantum systems on generic lattices. *New J. Phys.* **8**, 71 (2006)
10. N. Schuch, J.I. Cirac, M.M. Wolf, Quantum states on harmonic lattices. *Commun. Math. Phys.* **267**, 65 (2006)
11. K. Chen, S. Albeverio, S.-M. Fei, Entanglement of formation of bipartite quantum states. *Phys. Rev. Lett.* **95**(21), 210501 (2005)
12. G. Vidal, R.F. Werner, Computable measure of entanglement. Phys. Rev. **A 65**(3), 032314 (2002)
13. I. Peschel, Calculation of reduced density matrices from correlation functions. *J. Phys. A: Math. Gen.* **36**(8), L205 (2003)
14. Wolfram Research. http://functions.wolfram.com/01.28.27.0008.01
15. S. Dusuel, J. Vidal, Continuous unitary transformations and finite-size scaling exponents in the lipkin-meshkov-glick model. *Phys. Rev. B.* **71**, 224420 (2005)
16. C.M. Newman, L.S. Schulman, *J. Math. Phys.* **18**:23 (1977)

---

[3] The second and last term of Eq. K.10 cancel identically at this order

Curriculum Vitae

Hannu Christian Wichterich

Born November 12th, 1980 in Kiel, Germany. Study of physics at the universities of Konstanz and Osnabrück, Germany (Diplom 2007). Graduate student at University College London (UCL), United Kingdom, under supervision of Prof Sougato Bose (PhD 2010).

Since 2010 EPSRC PhD+ postdoctoral research fellow at UCL.

List of publications

1. HW, J. Vidal, and S. Bose. Universality of the negativity in the Lipkin-Meshkov-Glick model, Phys. Rev. **A** 81, 032311 (2010)
2. HW, J. Molina, and S. Bose. Scaling of entanglement between separated blocks in spin chains at criticality, Phys. Rev. **A** 80, 010304(R) (2009)
3. R. Steinigeweg, HW, and J. Gemmer. Density dynamics from current auto-correlations at finite time- and length-scales, EPL 88, 10004 (2009)
4. J. Molina, HW, V. E. Korepin, and S. Bose. Extraction of pure entangled states from many-body systems by distant local projections, Phys. Rev. **A** 79, 062310 (2009)
5. HW and S. Bose. Exploiting quench dynamics in spin chains for distant entanglement and quantum communication, Phys. Rev. **A** 79, 060302(R) (2009)
6. M. Michel, O. Hess, HW, and J. Gemmer. Transport in open spin chains: A Monte Carlo wave-function approach, Phys. Rev. **B** 77, 104303 (2008)
7. C. Mejia-Monasterio and HW. Heat transport in quantum spin chains: Stochastic baths vs quantum trajectories, Eur. Phys. J. Spec. Top. 151, 113 (2007)
8. HW, M. J. Henrich, H.-P. Breuer, J. Gemmer, and M. Michel. Modeling heat transport through completely positive maps, Phys. Rev. **E** 76, 031115 (2007)